Social and Technological Management in Dry Lands

Past and Present, Indigenous and Imposed

AAAS Selected Symposia Series

Published by Westview Press
5500 Central Avenue, Boulder, Colorado

for the

American Association for the Advancement of Science
1776 Massachusetts Ave., N.W., Washington, D.C.

Social and Technological Management in Dry Lands

Past and Present, Indigenous and Imposed

Edited by Nancie L. Gonzalez

AAAS Selected Symposium 10

AAAS Selected Symposia Series

Published in 1978 in the United States of America by

Westview Press, Inc.
5500 Central Avenue
Boulder, Colorado 80301
Frederick A. Praeger, Publisher and Editorial Director

Library of Congress Number: 77-93023
ISBN: 0-89158-438-2

Printed and bound in the United States of America

About the Book

Recent international attention has focused on desertification and its concomitants, especially the diminution of flora and fauna in arid and semiarid lands and the resulting reduction in the economic value of those lands. Natural factors such as drought and wind erosion, as well as various technological practices, have been blamed for the present situation in many countries. Most observers agree that human beings have been both perpetrators and victims of desertification.

Anthropologists have long been interested in documenting how different societies have affected and been affected by their environments. The papers in this volume present case studies of societies ranging from ancient Peru to contemporary Israel, along with several topically oriented works. All are designed to illustrate how various societies--whether by water management or by the exploitation of plants and animals--have attempted to achieve ecological balance. Social organization and ideology as well as technology are discussed as important variables affecting the ways in which populations adapt to, or cope with, desertification.

Contents

List of Figures

List of Tables

Foreword

The *AAAS Selected Symposia Series* was begun in 1977 to provide a means for more permanently recording and more widely disseminating some of the valuable material which is discussed at the AAAS Annual National Meetings. The volumes in this *Series* are based on symposia held at the Meetings which address topics of current and continuing significance, both within and among the sciences, and in the areas in which science and technology impact on public policy. The *Series* format is designed to provide for rapid dissemination of information, so the papers are not typeset but are reproduced directly from the camera copy submitted by the authors, without copy editing. The papers are reviewed and edited by the symposia organizers who then become the editors of the various volumes. Most papers published in this *Series* are original contributions which have not been previously published, although in some cases additional papers from other sources have been added by an editor to provide a more comprehensive view of a particular topic. Symposia may be reports of new research or reviews of established work, particularly work of an interdisciplinary nature, since the AAAS Annual Meeting typically embraces the full range of the sciences and their societal implications.

WILLIAM D. CAREY
Executive Officer
American Association for the Advancement of Science

About the Editor and Authors

William H. Bedoian is a fellow in human biology in the Department of Anthropology, Pennsylvania State University. He specializes in human biological and behavioral responses to ecological constraints in arid lands and in the applications of the human ecology approach to problems of developing nations in arid regions.

John W. Bennett is professor of Anthropology at Washington University, St. Louis, and visiting professor at the Land Tenure Center, University of Wisconsin-Madison. His interests are economic and ecological anthropology, with special applications to market agrarian economies. He has carried out extensive research in agricultural societies in North America, Japan, Taiwan, and India, including an ongoing, decade-long study of agricultural development and management under the environmental and economic constraints existing in Western Canada.

*Richard S. Felger is a research associate with the Arizona-Sonora Desert Museum, Tucson. He specializes in arid lands and new agricultural systems, and is an adjunct associate professor, Department of Ecology and Evolutionary Biology, University of Arizona. He is the author of numerous publications in these areas, including "Seri Indian Food Plants: Desert Subsistence Without Agriculture" (*Ecology of Food and Nutrition, *5(1), 1976).*

Nancie L. Gonzalez, vice-chancellor for academic affairs, University of Maryland, was formerly program director for Anthropology, Division of Behavioral & Neural Sciences, National Science Foundation. She is the author of A Heritage of Pride: The Spanish-Americans of New Mexico *(University of New Mexico Press, 1969) and* Saga of Santiago: Development & Dependency in the Dominican Republic *(Institute for the Study of Human Issues, in press).*

Susan H. Lees is associate professor of Anthropology at Hunter College, CUNY, and an associate with the Columbia University Seminar on Ecological Systems and Cultural Evolution. She is the author of numerous publications, including Sociopolitical Aspects of Canal Irrigation in the Valley of Oaxaca, Mexico *(University of Michigan, 1973).*

Michael E. Moseley, associate curator of Middle and South American Archaeology and Ethnology, Department of Anthropology, Field Museum of Natural History, Chicago, specializes in social patterns in Andean archeology and has recently published "Organizational Preadaptation to Irrigation: The Evolution of Early Water-Management Systems in Coastal Peru " in Irrigation's Impact on Society, *T. E. Downing and M. Gibson, eds. (University of Arizona Press, 1969).*

Gary P. Nabhan is a graduate research assistant in the Department of Plant Sciences, University of Arizona, and at the Arizona-Sonora Desert Mudeum. He specializes in agricultural ecology and arid land ethnobotany and is concerned with the conservation of rare desert crop plants grown by American Indians. He has published in the areas of domestication of desert plants and native American agriculture.

Sharon E. Nicholson is a postdoctoral fellow in the Advanced Study Program, National Center for Atmospheric Research, Boulder. Her area of specialization is historical climatology of Africa, and she has conducted research in Senegal and East Africa. She is the author of "Saharan Climates in Historic Times" in The Sahara and the Nile, *H. Faure and M.E.J. Williams, eds. (in press).*

Priscilla Copeland Reining is director of the Population and Desertification Project, Office of International Science, AAAS, and staff representative to the AAAS Committees on Arid Lands and on Desertification. Her specific areas of interest are land tenure, land use, kinship, and population in sub-Saharan Africa, both on site and using remote sensing data sources. She is author or coauthor of numerous publications, most recently Village Women: Their Changing Lives and Fertility *(AAAS, 1977).*

F. S. Vidal is associate professor of Sociology at the University of Texas, Arlington. He is interested primarily in the Middle East, culture change, and applied anthropology. His extensive field experience spans 21 years in Arabia and 7 years in Morocco.

Social and Technological Management in Dry Lands

Past and Present, Indigenous and Imposed

Introduction

Nancie L. Gonzalez

Anthropologists have studied the peoples and cultures of arid lands for over one hundred years, yet their findings are only slightly known outside their own field. Perhaps this is in part due to the fact that we have only recently begun to do problem-oriented research relating to aridity, drought and desertification. It is also the case that the ecological perspective in anthropology has been prominent for only the past 20-25 years. Although some environmental factors were dutifully recorded by early field workers, their analyses often failed to incorporate this dimension, or at best merely gave descriptive facts as part of the "setting". The natural environment, if considered as a dynamic factor at all, was most often viewed as having simply permissive or limiting effects on culture and its history. Still, since anthropology has always endeavored to understand the total sociocultural system, and since many of the world's peoples have inhabited arid and semi-arid lands for thousands of years, it was inevitable that their social, technological and ideological responses to and means of coping with such conditions should eventually become an important element of our research.

Julian Steward is most often credited with introducing and providing a theoretical framework for what has come to be known as cultural ecology.[1] He had predecessors, to be sure, and thinkers such as Daryll Forde in England and Clark Wissler in America had a clear effect upon his work. The approach has been largely wedded to evolutionary thinking in America, and has been most influential in so-called basic or "pure" research efforts by anthropologists. Applied anthropologists have more often utilized some form of functionalist approach in an effort to relate "beliefs and practices". This often led to a kind of "respect the *status quo*" stance on the part of the anthropologist. These papers are evidence that this is changing.

The present symposium grew out of an effort by the editor, as representative of the Society for Applied Anthropology to the AAAS, to bring together at the AAAS meetings in Denver a group of anthropologists who had worked in various arid or semi-arid lands and who also were concerned as social scientists with the problems of coping with drought and desertification.[2]

The reader may be surprised to note that the concern felt by most or all of the contributors is not the usual "How can we help the poor natives?" so much as it is, "How can we learn from examples of indigenous management techniques?" This also is a bit of a new thrust for applied anthropologists, whose employers have most often been those agencies of the politically dominant system charged with managing and improving the lot of a dependent or subject population. Aside from the borrowing of a few foods or medicines, however, it has not generally occurred to the developed countries that we could actually learn much of use to ourselves from societies we have thought of as inferior or "less developed". I believe this has been largely the result of the prevailing "bigger is better" ethnocentric philosophy of the industrial age.[3] Ironically, both the developed and the developing nations adopted this as a basic value, much to the consequent detriment of both. To be sure, the philosophy <u>seemed</u> to make sense for quite a while, especially as the countries which had developed earliest became richer and boasted higher standards of living, which others then wished to emulate. Certainly the technical superiority of fossil fuel-based sociocultural systems has been apparent in such things as medicine, transportation and machine production. It also seemed clear for agriculture--at least for a while.

Now however, the tide seems to have turned, and more and more thinking people in the "developed" nations are questioning the ideology of continued growth and the consequences in depleting natural resources. The fear that we may already have gone too far is the rallying cry of the environmentalists and those who would limit births on a worldwide scale. The search for "appropriate technology" takes us not only to the engineering drawing boards of our own culture, but to a reexamination of the more modest efforts of other societies in other times and places. It is natural that anthropologists should become involved in this new pursuit. It is also natural that we should turn our attention to agricultural practices, since the majority of those societies studied by anthropologists are directly dependent on the produce of the lands they occupy. We have collected massive amounts of data on varying types and systems of agriculture, much of which could be extraordinarily useful to future planners and developers. Unfortunately, most lies buried where only anthropologists can find it. Perhaps one day some concerned members of our field will retrieve and interpret these data for broader dissemination.

My own work in semi-arid lands has influenced my thinking on these subjects, and I'd like briefly to describe two cases which will serve to set the stage and to illustrate how much can be retrieved from basic anthropological studies not precisely directed toward problems of drought and desertification, nor even of agriculture *per se*.

In the 1960's I did ethnohistorical research on the Spanish-American population of the Southwest--particularly in what is now New Mexico (Gonzalez 1969). These early settlers could not have achieved agricultural success without a sophisticated irrigation system which controlled the use of stream waters so that each farmer received his share. Some of the technology involved was learned from local Pueblo Indians, with improvements brought from Spain. Social organization was equally important in management, and the "ditch associations", as they were called, survive in some of the Hispano rural communities to the present day. Although this society did not achieve the massive and impressive engineering feats described by Moseley for ancient Peru, its members nevertheless exhibited a thorough understanding of what was needed to build, maintain and manage a hydraulic system appropriate to their needs. No overall State control was present or necessary, given the conditions then existing--a relatively small population size and a minimally sufficient water supply.

However, as the population grew, so did the herds of sheep which these mountain farmers came to depend upon more and more in the early 19th century. Heavy grazing tended to alter runoff patterns, which affected areas hundreds of miles to the south. With the conquest of the region by the United States in 1846 came an even heavier population influx, and competition and hostilities between Hispano sheepherders and Anglo cattlemen became severe. With the introduction of agriculture on a massive scale in the southern part of the state, the need for water and range management at a political level higher than the community became apparent.

Elephant Butte Dam, built by the U.S. government in 1917, helped control alternate flooding and drought. In the 1930's herd sizes were controlled by law, much to the distress of thousands of small Hispano and Native American farmer-herders. A system which had worked well for centuries was simply obsolete under the new conditions. Many were driven to the towns and cities or out of the area completely, and the economic base of New Mexico came to depend less on agriculture and more on tourism, health and retirement, as well as military and other government installations.

Throughout the transition period there was a tendency, which continues today, to blame the Spanish and Native Americans for what was believed to be range mismanagement, with subsequent degradation of the land. However, we should not forget that a perfectly satisfactory balance had once been achieved and that it was largely the intrusion of a foreign policy and population which turned the ecological tables. Some internal adjustments most likely would have been required to accommodate the growing indigenous population, but they might have been accomplished with less trauma for both land and people.

My second encounter with semi-aridity and its implications for humans was in the Dominican Republic, where I carried out field research in 1967-68 with funds from the National Science Foundation. I concentrated on the northern valley known as the Cibao, sometimes referred to as "the breadbasket of the nation" (see Gonzalez 1977). Although the Indians living there at the time of Columbus and the early settlers had both practiced fairly intensive agriculture, it was not until the last decade of the 19th century that irrigation works were introduced. These allowed the opening of some new lands, but more importantly, made it possible to grow new crops--especially wet rice.

The new agriculture was dominated by a class of merchants turned agrobusinessmen, who were also powerful figures in regional and national politics. The canal system was supposedly controlled by the government, but there were breakdowns both in maintenance of the channels and in payment of water taxes by users. Furthermore, the system was inadequate in that it did not provide for proper drainage. In about 70 years much of the land had deteriorated--largely through increasing salinity--to the point that the large owners began to sell off enormous tracts--often to the government, which then redistributed it in smaller plots through the agrarian reform program.

In other publications (Gonzalez 1973; 1977) I have described these and related events in greater detail. In contrast to the New Mexican case, it can be demonstrated beyond a shadow of a doubt that the land became degraded through misuse rather than overuse. Here, however, it is not the small, poor, indigenous farmer using time honored folkways to exploit the land, but the representatives of "modernity", armed with both capital and supposedly scientific know-how to whom the accusatory finger must point. Worst of all, the poor management techniques are being perpetuated by those who now farm the same degraded lands. They do not understand why their yields are low when those who preceded them became rich. Efforts are being made by agricultural scientists to develop new crops which will resist salinity, but there is little evidence that the small farmers are learning to take advantage of these. Irrigation with no or poor drainage continues, to the increasing jeopardy of the land.

In addition to the problems related to salinity, there is also an increasing dependency upon chemical fertilizers in an effort to improve the yields, which fluctuate enough from harvest to harvest to give the small farmer hope that he can indeed control matters if he but adopts more "modern" techniques. My notes contain many descriptions of crops burned and destroyed by the application of too much fertilizer at the wrong time. Furthermore, the streams which receive some runoff are increasingly polluted by these practices, which further alters the ecological balance in the area.

If we combine such examples with those involving worsening climatic conditions in certain specified areas, we must begin to worry about food production in the world as a whole.

Lest some accuse us of wearing the rose-colored glasses of romanticism or of worshipping the noble savage, let me point out that the papers in this book illustrate both the successes and the failures of indigenous systems. As Reining's and Nicholson's discussions note, there have always been both natural and political hazards which must be overcome if an otherwise successful management program is to survive. Bennett reminds us that an elaboration of a system requires a recognition of all its parts, including the social and ideological, as well as the technological. Too often in the recent past it has been precisely the introduction of Western values, whether at elite or grass-roots levels, which has brought about disaster, as illustrated in the papers by Vidal, Bedoian, Lees, and Felger. In some cases indigenous peoples have not survived, or they have abandoned their homes and migrated to new lands and new lives. This process undoubtedly occurred often in antiquity, as suggested by Moseley's paper, and continues to the present day.

Taken together, these papers provide illustrations from many, though not all thw world's arid and semi-arid zones. Moseley and Lees deal with hydraulic management in several agricultural systems, ranging from ancient Peru to modern Israel. Vidal treats the delecate manipulation of scarce surface water by Bedouin herders in the extremely arid Arabian peninsular--a nearly lost technique.

Bedoian's paper contrasts indigenous with more modern and experimental agricultural and herding practices on the Saharan fringe in southern Tunisia, while Felger and Bennett use examples from the United States. My introductory notes to the papers will provide further contextual remarks to help guide the non-professional reader toward an understanding of the kinds of contributions anthropologists can make.

Though we recognize that much is still to be said and done, we hope these papers will have some small impact on both our fellow anthropologists and upon other scientists and administrators concerned with the human factors in dry lands management.

Notes

[1] See Harris 1968: 654 ff. for an informative discussion of the development of Steward's thought.

[2] All the papers published here were originally invited and coordinated by me, to be sponsored by Section H and SfAA. The program chairman for AAAS requested that our symposium be combined with another, organized by Priscilla Reining and Anthony Leeds and sponsored by the Office of International Science of AAAS and the Society for the History of Technology. Consequently, there were other titles included on the written program under this session. However, of these others, only a paper by Leeds was actually delivered in Denver. That, plus another given by Theodore Downing as a part of the SfAA contribution, were not finished in time to be included here. We regret their omission and look forward to their eventual publication elsewhere.

[3] The fact that it is not inherently related to conquest _per se_ is shown by the many invaders in the past who adopted most or all of the conquered sociocultural system: witness the case of the Mongols and the Chinese, and that of the Aztecs in the Valley of Mexico, among others.

References

Gonzalez, Nancie L.
1969 _A Heritage of Pride: The Spanish-Americans of New Mexico_. Albuquerque, University of New Mexico Press.

Gonzalez, Nancie L.
1973 "The Sociology of a Dam". _Human Organization_, 32(1): 353-360.

Gonzalez, Nancie L.
1977 (in press) _The Saga of Santiago: Development and Dependency in the Dominican Republic_. Philadelphia, Institute for the Study of Human Issues.

Harris, Marvin.
1968 _The Rise of Anthropological Theory_. New York, Thomas Crowell.

The technological solutions of preindustrial societies are worth examining, especially when research suggests that they have achieved considerable success in modifying or controlling the environment. We are too accustomed to assuming that "modern" science and engineering are inevitably superior, both to prehistoric efforts and to those of more contemporary "traditional" peoples. Moseley's archeological perspective should give the reader pause, for he concludes that the ancient Peruvians had a highly sophisticated knowledge of hydraulic engineering which in some ways rivaled our own. Even though their system ultimately failed, he warns that we should try to learn more about it in order to avoid the same fate if possible.

1

An Empirical Approach to Prehistoric Agrarian Collapse: The Case of the Moche Valley, Peru

Michael E. Moseley

The world's increasing population confronts us all with the pressing problem of how to feed ourselves and our ever-increasing brethren. One solution is to transform deserts that now cover one-eighth of all land surfaces into fertile food-producing regions by means of irrigation agriculture. Yet reclaiming deserts is complex, entailing the creation of symbiotic relationships delicately balancing interactions of plants, land, water and people.

The formerly successful transformation of arid lands in coastal Peru into fertile fields is the principal theme of this article. However, my concern is also to generate an awareness that contemporary irrigation programs can profitably draw upon archaeological research.

Today large-scale irrigation projects are corporate undertakings generally instigated at state or national levels. Following assessment of water and land availability, costs and hypothetical returns, the combined forces of modern science and technology are marshaled to construct a canal delivery system and implement farming. True fruition of such reclamation projects comes when caloric out-put ultimately exceeds the original investment input, and this generally requires decades.

Most aspects of modern science and technology matured in temperate rather than arid climates. Thus, contemporary irrigation engineering that is exported from first world marketers to third world consumers cannot be said to have developed in either a social or physical context traditionally concerned with or adjusted to desert life. Therefore, if decades are needed to assess the success of modern reclamation projects, then what is being marketed is largely an untested product.

In some cases nations exporting irrigation know-how have applied this skill to their own needs. These constitute test cases, and perhaps the most grandiose is the Colorado River system in the southwestern United States. If this system were to end today would future archaeologists judge it as having been successful? Probably not, if agricultural output were the sole criterion of judgement. Diverting great quantities of water to urban areas, such as Los Angeles, would be deemed wasteful of a food producing resource. Mismanagement leading to salinization of vast tracts of agricultural land near the river mouth would be alleged because these farm lands were permanently lost to future generations.

To say that many modern reclamation projects proceed on a trial-and-error basis would be charitable, but wrong. Most large programs have not been operational long enough for the errors to materialize. Thus, the disquieting question is: "Are such projects actually proceeding on a hit-or-miss basis so far as long term consequences are concerned?"

An alternative to this scenario would see modern irrigation endeavors drawing understanding of their long-range consequences from the study of previous hydraulic projects, especially those operational for centuries with proven productivity, and whose impact on the social and physical environment is known. An optimal situation would be one in which a prehistoric project brought more land under cultivation than is presently farmed in the same area with the benefit of modern technology. The two water management systems could then be compared and contrasted, and factors underlying the demise of the earlier, more productive system could be isolated. This "empirical" approach offers the potential of removing the long-term consequences of irrigation from the sphere of mere speculation and raising them to the realm of known probabilities.

Situations as described above do not exist within the physical boundaries of countries exporting irrigation technology today, which is probably why the approach just outlined has not received serious attention. Nonetheless, cases of very old, highly successful irrigation programs do exist elsewhere in the world, and their study is of great relevance.

In the Moche Valley, on the northern coast of Peru, pre-Hispanic populations irrigated 30 to 40% more land than is reclaimed in the same region today. Before discussing the rise and fall of this ancient water management system I will briefly review its broader physical, social, and technological setting.

The Desert

Flanking the western slopes of the Andes is the driest coastal desert in the world. Annual precipitation along the northern coast averages between about 0.5 and 1.4 cm. and as a result the area is bleak, with little or no native vegetation. However, the climate is remarkably mild with minor variation in temperature from summer to winter when coastal fogs set in (Robinson 1964; McGinnies et al. 1968).

The Peruvian desert is crossed by a series of 57 water courses that have affluents reaching into the mountains where annual rains fall at elevations above about 1,800 m. Runoff cascades down the western face of the Andes in steep, narrow gorges. Nearing the coast gradients decrease, and flood plains fan out before the water courses discharge into the Pacific Ocean. The streams and rivers are spaced roughly 40 km. apart. The southern mountains receive less precipitation than their northern counterparts; therefore southern streams tend to discharge substantially less water than the northern rivers.

For more than three millenia economic adaptation to the desert has focused on these streams and rivers as water sources for irrigation agriculture. However, water is but one variable in the symbiosis that underlies successful cultivation. Land is another critical factor. Along the central coast of Peru the Andes and their foothills push out to the Pacific, and flat arable land is at a premium. This often deprives the people from gaining the full economic potential of large rivers, and in such settings it is the scarcity of arable land, rather than water, that sets limits on agriculture. In the south there is a coastal plain, but runoff is less and the availability of water rather than land sets tight strictures on agricultural out-put. There are several large southern rivers, but these lie deeply entrenched below the coastal plain. Here the physical separation of abundant land and water curtails large-scale irrigation.

The most auspicious combination of land and water lies in the north, where relatively well watered mountains stand back from the coast. At the mouth of the Rio Moche a true coastal plain emerges and becomes progressively wider to the north, where it is crossed by three very substantial rivers (the Chicama, Jequetepeque, and Lambayeque-Chillon) as well as by smaller streams.

The Rio Moche is a medium sized drainage basin of about 1,500 km^2. Starting with affluents at the continental divide more than 4,000 m. above sea level, the river rapidly

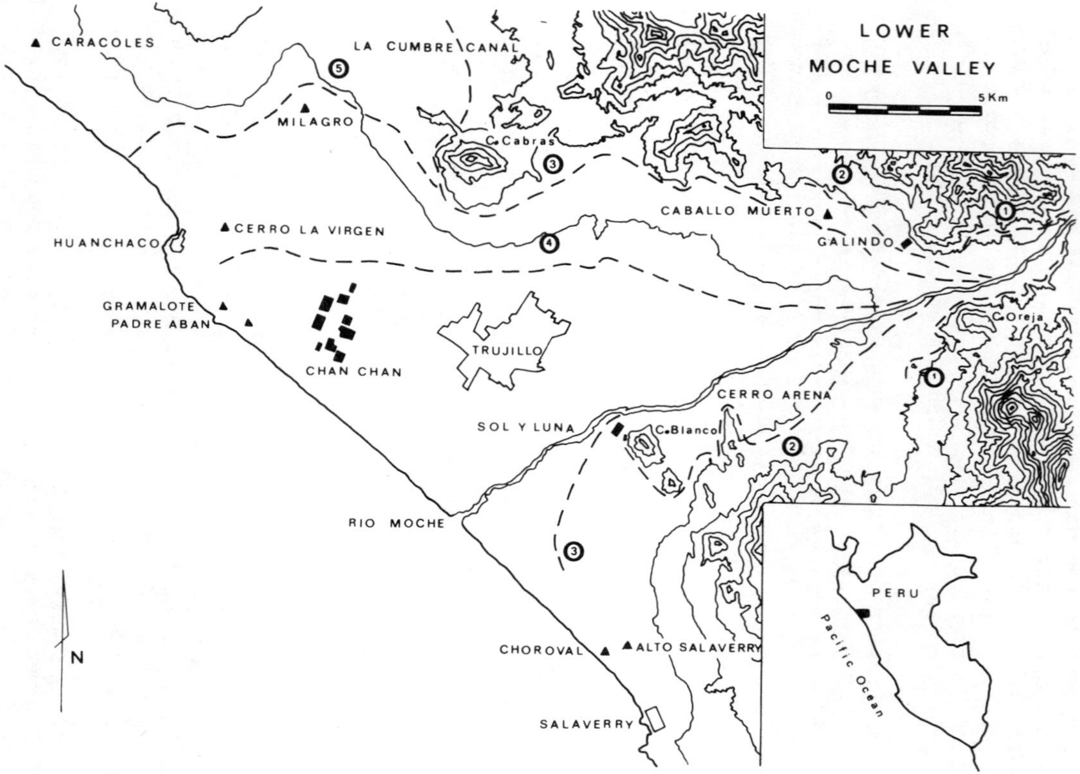

LOWER
MOCHE VALLEY
0
5Km
CARACOLES
LA CUMBRE CANAL
MILAGRO
C.Cabras
CABALLO MUERTO
GALINDO
CERRO LA VIRGEN
HUANCHACO
GRAMALOTE
PADRE ABAN
CHAN CHAN
TRUJILLO
C.Oreja
CERRO ARENA
SOL Y LUNA
C.Blanco
RIO MOCHE
CHOROVAL
ALTO SALAVERRY
SALAVERRY
PERU
Pacific Ocean
N

Figure 1.

descends the Andes in a narrow, incised canyon. Nearing the coastal flats the flood plain fans out to form a broad alluvial triangle about 25 km. wide at the valley mouth. This lower region has both the largest canals and the greatest amount of arable land in the drainage area. During the Pleistocene the river course slowly shifted southward, leaving more flat land on the north than on the south side of the valley. Three large rock hills (Cerros Orejas, Arena, and Blanco) flank the south side of the river. The only northern hill (Cerro Cabras) affecting the lateral expansion of canals is more than 10 km. away from the water course. This asymmetrical configuration of the lower valley endows the north with more agricultural potential than the south (Figure 1).

The Moche and other valleys to its north are among the most productive lands in the world for sugar cane, their principal contemporary crop. The region also yields returns on maize farming substantially above world averages. The favorable combination of sunshine, mild but constant temperatures, and good soils makes this area one of the most productive agricultural settings in the New World.

Not all reclaimed land in the lower Moche basin can be farmed year around because of scarcity of water at certain times of the year. The river fluctuates seasonally from a high of 10 to 34 m^3/second between January and May, to less than four m^3/second in remaining months. Water rights tend to be such that up-stream canals have priority over their down-stream counterparts. Thus, fields dependent on the latter are more affected by seasonal lows in the river than are those further in from the coast. Some lands near the coast are low lying and poorly drained, whereas inland soils are not. The interior combination of good water-rights and favorable soils has resulted in a situation where now almost all land from the valley neck down through the central valley is owned and farmed by a single large plantation specialized in mechanized cultivation of sugarcane. What may be termed the "high risk" areas, those closer to the coast or on the tail ends of long canals, are held by small farmers tending crops for local consumption. This pattern of centralized ownership and working of low risk interior lands accompanied by decentralized holdings and small-scale farming of less secure, and generally more seaward-terrain, is common in other valleys of the region.

Environmental Change

The physical environment along the Peruvian coast is extremely stable in many respects. From a seasonal perspective there are winter fogs and fluctuations in river discharge.

From a year to year perspective there is some variation in highland rainfall, and occasionally droughts affect the western watershed. From the perspective of geologic time, climatic and physical conditions are thought to have been relatively constant for at least four or five thousand years (Craig and Psuty 1968).

From an archaeological perspective it is, however, possible to isolate at least three variables that have affected human desert adaptation at various times. Contemporary populations are, apparently, aware of only one of these variables, the <u>niño</u> phenomenon. Niños entail short-term shifts in meteorological patterns associated with temporary but marked changes in the complex of ocean currents that flow parallel to the Andean coast. There is considerable variation in the intensity of these perturbations. Generally, niños do not last very long--rarely more than a month--nor do they strike very often. Occurrences are thought to be cyclical, with appearances variously estimated at every five, seven, or more years. Niños are more intense in the north than in the south, and near the Peruvian-Ecuadorian border they are both more common and of greater impact than further south. The Moche Valley region has experienced niños during the present decade, but the last occurrence with major impact was in 1925. This was a staggering cataclysm. The temporary changes in sea currents resulted in massive loss of marine fauna. Meteorological shifts resulted in torrential rains and extensive flooding of coastal lands. Rains accompanying both large and small niños produce flash flooding and the washout of roads, canals, and adobe houses in which most people live. It is little wonder then that people are aware of the niño phenomenon, infrequent though it may be, or that the 1925 cataclysm is still talked about.

Two environmental variables about which there is little awareness are dune formation and tectonic uplift of the coast line. Saltating sands and active dunes are commonplace in the Andean desert. Sand often accumulates to depths of hundreds of meters on windward and leeward hill-sides, creating very impressive scenery. There is a tendency to regard the sand accumulations as inherent, primordial aspects of the landscape because the inception of dune formations and their subsequent movements are difficult to date. However, in some valleys there are prehistoric sites and settlement areas now swamped by active dunes, and it is evident that these areas would have been uninhabitable if the saltating sands had been present during the original human occupation. These situations have still to be systematically studied, and the affected settlements themselves are largely undated. It is unclear if the sites were all swamped at the same period, or if burial occurred at different times in different locations. The latter would suggest that dune formations were localized, sporadic events. The former would imply a widespread process affecting much of the coast at the same general point in time. Whatever the case, it is clear that dune formation altered the physical environment after the onset of irrigation agriculture.

The Andes are a young and geologically unstable range. Separate archaeological studies scattered along more than 1,000 km of coast--from Talara (Richardson 1973; 1974) in the north to Paracas (Craig and Psuty 1968) in the south--all point to tectonic uplift of the shoreline during the last four to five thousand years. Around 2,000 B.C., before the practice of intensive irrigation, a far reaching maritime adaptation supported large populations concentrated in fishing communities strung along the coast. Many maritime settlements were situated atop wave-cut cliffs and beach lines of protected bays that afforded favorable conditions for netting small fish, such as anchovies, and collecting mollusks and other sea foods. Around 1,800 B.C. a majority of these communities were abandoned, and today many of the formerly productive bays are dry sand flats (Moseley 1972). In some cases, successions of receding beach lines are occupied by a series of progressively more recent archaeological sites marching seaward in slow pursuit of the gradually retreating waters (Richardson 1974; Engel 1957; Craig and Psuty 1968; 1971).

While the archaeological record is explicit about a general regime of uplift, the possible unity of this process either in terms of rates of elevation or areas simultaneously affected remains unfathomed. In theory, at least, bay-drying may have contributed to abandonment of early maritime communities. Exposure of vast sand flats may have provided the raw materials required for dune formation. Finally, uplift may have contributed to entrenchment of coastal rivers, and in some cases the loss of irrigation potential.

If the impact of niños, dunes, and uplift has been little explored, it is because these environmental features have only recently been identified by archaeologists moving out of the traditional concerns of their discipline and into the field of irrigation. Unfortunately, these factors generally remain unacknowledged by engineers and technicians who build canals across dry washes that will flood, along the base of dune formations that will move, fed by rivers that will change courses. Ignoring these sporadic and slow moving processes of environmental change is not conducive to successful reclamation.

Canal Strategy

Indigenous populations pursued multiple agricultural strategies in adjusting to the stable and unstable aspects of their physical environment. Irrigation based on river-fed canal delivery has been the dominant mode of farming for millenia, and contemporary irrigation systems rely extensively on the continued use of pre-Hispanic canals. The general characteristics of native water delivery systems will be discussed before considering complementary strategies.

Valley topography exerted strong influence on the configuration of local canal systems. Most valleys are "U" shapped in profile, becoming progressively wider toward their mouths. Canals are led off a river laterally at an angle so that the channel slope is less than the gradient of the river. Seen from the air, an ancient irrigation complex has a configuration similar to a series of nested "Vs" (V V V). The point of each V represents the intake of two canals at the river. The diverging arms of the V reflect the course of the canals as they spread apart from the river and reach out into flatter lands nearer the coast (cf. Figure 1).

In an indigenous system agrarian expansion was dependent upon reclaiming new terrain by three principal means. First, the capacity of an extant canal could be enlarged, allowing more water to be carried. Second, an existing canal could be lengthened, pushing it into new land. And third, a new channel could be cut at an elevation above an extant canal, either by starting it further upstream, or by cutting it with a lower gradient so it could feed off the existing canal. These three principles--"capacity enlargement", "distal extension" and "elevational increase" -- were used individually and in combination to exploit progressively more land through time.

For most of the 57 drainages crossing the Peruvian coast, limitations in water or land availability set strictures on the scope of reclamation, such that each river and its lateral canals formed an individual complex separated from the next by a stretch of desert. In five cases, favorable combinations of water, land, and low topographic relief allowed canal link-ups of two or more valleys.

Three megasystems are still operational. Lima is located in a double-valley megasystem. There is another 160 km. to the south, as well as a three-valley complex 170 km. north of the capital city. The Moche drainage was part of a two-valley complex that received water from the Rio Chicama by means of the La Cumbre canal, which was more than

80 km. long. Immediately north lay the largest irrigation system on the Andean coast, the Lambayeque complex. Here five drainages from the Rio Jequetepeque to the Rio Motupe were interconnected by canals integrating more than 100 km. of coastal plain. The Lambayeque megasystem serviced one-third of all agricultural land along the Peruvian desert (Kosok 1965). As in the Moche-Chicama area, the major canals connecting the five northern valleys have been out of use for centuries.

One of the largest contemporary reclamation programs on the coast is the Tinajones-Taymi project entailing an attempt to use modern technology to reactivate one of the intervalley canals of the Lambayeque megasystem. This major undertaking relies largely on foreign engineering and financing. As is characteristic of most modern reclamation endeavors, none of the project scientists or engineers has tried to isolate the reason that the canal being reactivated fell into disuse in the first place! From an archaeological perspective the Tinajones-Taymi undertaking is mute testimony to the anti-empirical bias of nations exporting irrigation technology to-day.

Complementary Strategies

One reason that there is less land under cultivation in the present than in the past is because a greater variety of agricultural strategies were systematically pursued by pre-historic populations. These strategies developed in response to a set of variables that included water table level, drain-age conditions, and water-borne sediments and minerals.

The Andes are not heavily forested, so erosion charges the coastal rivers with large loads of sediments and miner-als. This has at least four irrigation-related consequences. First, to avoid excessive channel clogging, canals must be designed so that as much of the sediment load as possible will be kept in suspension until the waters reach the fields. Second, through time, substantial sediment deposits build up in fields and along canal courses, gradually transforming the natural topography into an irrigation shaped landscape in which slopes and contours conform to the distribution and configuration of individual fields and canals. The process is most pronounced near rivers where water management systems carry their greatest silt loads. Here major canals often ride along the edges of irrigation-created terraces that are four or more meters high. With time the depositional process tends to implant a canal system, and makes radical changes in the system progressively more difficult to execute because very large labor investments are needed to rework the irrigation-created landscape. Third, a positive feature of suspended sediments is that they contribute to soil formation and field fertilization. Many sections of the desert initially had very poorly developed soils, and the depositional processes accompanying their irrigation gradually led to the formation of better agricultural matrices. Fourth, a negative feature of irrigating with Andean runoff is a propensity for salinization to occur when drainage conditions are poor or when there is insufficient water to keep the accumulating minerals and carbonates flushed from agricultural lands. High water table conditions exist along many rivers, particularly near delta regions and along the coast line. There, poor drainage can lead to permanent land loss due to salinization. Thus, irrigation is least auspicious in these settings, and most propitious in higher lands along valley flanks.

Agricultural strategies complementing river-fed irrigation seem to be both more numerous and more elaborate in valleys with prevalant high water table conditions, such as the Viru drainage to the south of the Rio Moche. Studies in Viru by Michael West (1977) have greatly expanded the known repertoire of indigenous farming techniques. Combining this information with data from other northern drainages provides a composite index of cultivation strategies used in the past.

1) Flood-water farming. When coastal rivers reach their seasonal crest, sections of flood plain are inundated, soaked, and then left exposed. Sowing of crops can then take place in these "self-watered" lands without investing substantial labor in manipulating either water or land. Farming of this type is by nature seasonal and the limited availability of river-flooded lands constricts production. However, little labor is involved, and this mode of cultivation is inferred to have been important during the inception of coastal agriculture before the advent of large scale canal irrigation (see below).

2) Ground water farming. In some situations the water table lies close to the ground surface, and planting can take place with little effort. This is another example of self-watered lands. Their exploitation may be less seasonal than flood-water farming, but the quantity of such terrain is relatively limited. This agricultural strategy finds scattered current use in the lower Viru Valley, and it may have been important at an early date before the advent of irrigation.

3) River-fed irrigation. This is the dominant agricultural technology on the coast, and with large, perennial

rivers cultivation can go on year-round. Water is the primary variable being manipulated, and the construction of canal delivery systems is highly labor demanding.

4) Ground water irrigation. Surface accumulations of ground water are common in low areas of many valleys. Where such sources occur sufficiently far inland they can be used to feed canal systems that move water laterally and down-valley into areas with better drainage. The technique is presently employed in both the Viru and Moche valleys. Water is the manipulated variable, and the size of dependent canal systems, and thus the prerequisite labor investment, seems to rest on factors such as water quantity and topographic accessability of soils with suitable drainage. This technique is presumed to have pre-Hispanic origins, but has yet to be demonstrated archaeologically.

5) Drained land cultivation. Areas with high water tables or surface accumulations have been drained by cutting trenches or sangrias which "bleed" off excess moisture so as to create workable agricultural matrices. In favorable situations the water can be directed into source canals; often, however, it is simply drained into lower areas, a river, or the ocean.

6) Ridged field cultivation. An alternative approach to conditions of excessive moisture involved altering the land surface. This was done by mounding up earth, creating high ridges separated by furrows capable of serving as drains. The ridges then served as elevated planting surfaces. Ridged fields involve a relatively high initial labor investment, but once operational such systems are presumed to have been stable for long periods. This farming technique was used in widely scattered areas of South America (Parsons and Denevan 1967), but in coastal Peru ridged fields have only been found in the Casma Valley.

7) Sunken field agriculture. A common strategy involving land manipulation entailed the creation of farm plots by excavating a planting surface down to a depth close enough to the water table so that naturally occurring moisture was available for plant growth. Depending on ground water depth and plot size, creating a sunken field can be very labor demanding. However, it is a relatively efficient means of exploiting certain areas where irrigation is impractical due to a lack of surface water, or where salinization presents a potential problem.

Listing these strategies here is intended to make their existence known, and does not imply thorough archaeological understanding of the phenomena. In some cases sunken fields were partial solutions to an insufficient supply of surface water for canal farming. However, the majority of non-irrigation strategies seem to be responses to problems of high water table, poor drainage, and salinization. It is not clear how successful these responses were over the centuries. A study of one set of sunken fields revealed that calcium-rich hardpans had apparently formed at the old planting surfaces, presumably decreasing agricultural productivity through time (Parsons and Psuty 1975). Today some farmers simply cut out these formations, making their fields somewhat deeper, thus minimizing loss of productivity. Still, little is known about the problems or productivity of techniques such as ridged fields, drained land, and ground water farming.

It is probably a valid generalization to say that the basic indigenous strategy consisted of applying the most productive technique to the best available land, which entailed the use of river-fed irrigation on valley sides and lateral sections of desert. Secondary techniques were then used for low lying and coastline lands. Yet, in pondering prehistoric land and water management, two things are worth remembering. First, irrigation itself raised the water table and probably accented seasonal fluctuations, thereby generating a new set of problems. Second, the entire coast was, and is, experiencing uplift, thus creating another set of problems which probably did not cancel out the first set.

Social Setting

In the Moche Valley, irrigation agriculture is inferred to have begun by 1,500 B.C. However, relatively little is known about the earliest reclamation activities, or the social context in which they took place. The archaeological record assumes sharper focus around the time of Christ, and for the last 1,500 years of pre-Columbian occupation there is a better historical understanding of the region.

By about 1,000 A.D. the valley population had evolved into two caste-like groups, the nobility and the commoners, each of which was internally stratified. The great majority of people were commoners, and different economic pursuits correlated with different class distinctions. Most of the populace engaged in farming and constituted a low-ranking class of agricultural peasantry. Fishing was a specialized activity and fishermen formed a distinct social group. The highest ranking but smallest class of commoners was made up of artisans and skilled craftsmen, who were concentrated in and around major urban centers. They formed the urban proletariat, whereas other commoners generally resided in rural settings (Moseley 1975a).

Membership in the nobility was determined primarily by kinship, making this a closed and hereditary social category. There was a complex system of internal ranking. At the apex of the hierarchy stood a king or emperor who was held to be divine or semi-divine. His offspring formed a high ranking lineage, and the immediate descendants of each former emperor comprised separate royal lineages. Below the royalty there was an aggregate of lesser nobles. Presumably this included the issue of the emperors' concubines as well as the former rulers and aristocracy of conquered provinces (Moseley 1975a).

The noble caste was basically an urban elite. The formal apsects of the political, economic, and religious systems were controlled by these people. Thus, the nobility formed the bureaucracy that organized corporate affairs and directed the activities of the commoners. A preoccupation with theft pervaded the legal system, and the nobility seems to have regarded property ownership as their exclusive prerogative by divine right (Rowe 1948). In overview, the sociopolitical system was highly centralized and totalitarian, with some benevolent tempering. Exactly how and when this system developed is not clear, but it seems to have been fairly well crystalized by the time of Christ.

The economy was agrarian, but the peasantry did not own the lands they worked. Instead, corporate proprietorship prevailed. In the Inca Empire there was a threefold division of land. In a given area roughly one third of the land belonged to the local community as a corporate entity, and members were assigned plots for personal use. Another third of the land was held by the state church or religion, and the remaining third belonged to the emperor. These holdings were also worked by the peasantry of the local community. In the Moche Valley corporate proprietorship was the norm in pre-Inca times, but it is not clear if a threefold division of holdings also existed. Most land was state owned as well as centrally organized and operated. However, the exact manner in which proceeds were distributed among community, religious, and political interests is unestablished.

Andean societies did not employ currency or monetary systems. Instead, economic and political institutions rested upon the practice of extracting taxes in the form of labor. The Inca expression of this practice, the <u>mit'a</u>, was described by early Spanish chroniclers. This tax required men to devote periods of service to state and corporate interests, while women produced quantities of fabrics for the administrative hierarchy.

The operation of mit'a-like principles has a number of physical expressions that have received archaeological attention in the Moche Valley. Large scale construction projects, such as platform mounds or long walls, were subdivided into repetitive segments or modules. Some projects employed mud bricks with "makers' marks" or distinct symbols impressed on their surfaces. In such projects, each segment or module is built entirely of bricks with the same symbol, while bricks in an adjacent segment will all have another, different makers' mark. This phenomenon is thought to reflect three things. First, segmentation is an expression of separate labor parties working at individual, but repetitive tasks. Second the association of different segments with different makers' marks identifies specific work parties who not only produced their own bricks, but transported them to the construction site and were then responsible for laying the abodes in a particular segment. And, third, different makers' marks distinguish different social groups given separate status for taxation purposes, and segments were the work units or measures of labor by which these different groups paid their tax obligations (Moseley 1975b).

This organizational principle is also evident in canal construction. The La Cumbre Canal, which transported Rio Chicama water into the Moche drainage, had to cross a wide intervalley divide. Labor parties working from both the Moçhe and Chicama sought to connect two legs of the canal just north of the divide. Here a rocky knoll made direct line-of-sight surveying between the two sections impossible. As a result, at least three abortive linkups were started before achieving a channel course with the proper gradient. Two attempted linkups were about 1 km. in length, while the third was somewhat shorter. Resurveying led to recognition of the engineering errors before any of the abortive canals were fully finished. All three attempts exhibit the segmentation phenomenon in the form of completed canal sections 50 to 70 m. in length interspersed with unexcavated segments. The latter were often delineated by lines of stones along each side of the projected water course (Kus 1972; Moseley 1975b).

Another agrarian related feature concerns the administration of work forces mobilized by labor taxation. Certain rural sites can be identified as formal, state-built complexes. The architectural canons at these sites imitate features of buildings at the capital of the state to which the complexes were subservient. These rural sites lacked a lower class resident population, and thus were quite isolated. However, they always occur in areas of ancient fields or adjacent to major canals that were corporate construction projects. This co-occurence is considered to be a functional association in which the state-built complexes served as administrative foci for directing labor forces engaged in building or operating agricultural works (Keatinge 1974).

Labor taxation has a number of important agricultural correlates and consequences. Economic growth in an Andean state was basically a matter of increasing control of agrarian output. To this end taxes could be invested in two ways: the labor mobilized by the mit'a could be used militarily in the conquest of additional lands or it could be invested in the reclamation of new lands. In an absolute sense the former increased control while the latter increased output. The second strategy underlies the vast irrigation projects carried out on the coast, as well as the Incas' later terracing programs in the Andes. If a state invests in reclamation its ensuing ownership of the new agricultural area is not surprising. Corporate proprietorship, centralized administration of arable land, and labor taxation were thus functionally interrelated institutions. In many ways this social matrix formed the foundations for making ancient reclamation programs more embracive and successful than their modern counterparts.

The Moche Valley Agricultural System

Three strategies or techniques were used by indigenous populations to push Moche Valley reclamation to its maximum: river-fed irrigation, ground water irrigation, and sunken fields. Sunken fields occupied a roughly triangular area in the center of the valley mouth stretching east from Chan Chan to what became the colonial city of Trujillo, and then down onto the coastal flats south of the river mouth. A short canal system fed by ground water paralleled the north side of the river between the flood plain and Trujillo. These secondary strategies are associated with low lying lands, a high water table, and poorly drained soils.

Most reclamation was carried out by river-fed irrigation. In overview there are, or have been, eight major canals in the lower valley (Figure 1). Three are on the south side of the river and five are on the north side. These are generally "Maximum Elevation Canals" (MECs), which have the highest elevation of all canals in a particular section of the valley. Because MECs run along the valley sides and feed water downslope toward the valley mouth, they represent the

TABLE I. AREAS OF RECLAIMED LAND:

PRE-HISPANIC AND MODERN

VALLEY AREA	CULTIVATED LAND IN SQUARE KILOMETERS
A. North Side:	
1. Modern:	103.041
2. Pre-Hispanic:	175.531
3. Pre-Hispanic not now in cultivation:	70.491
B. South Side:	
1. Modern	25.295
2. Pre-Hispanic:	37.841
3. Pre-Hispanic not now in cultivation:	12.548
C. Whole Valley:	
1. Modern:	128.333
2. Pre-Hispanic:	211.372

maximum limits of reclamation. Because the valley is asymmetrical in profile and flatter on the north than the south, northern MECs are generally longer than their southern counterparts.

Areas in the western part of the valley also received Rio Chicama water delivered by the La Cumbre intervalley canal. This channel was brought into the north side of the valley, and above Chan Chan united with the Vinchansao canal, the largest of the MECs. The combined waters were then carried westward irrigating three plains that make up the northern flats of the valley: the Pampas Esperanza, Rio Seco, and Huanchaco.

The following summary lists the principal canals in the Moche Valley system. For ease of reference the channels are given letter-number designations, and modern counterpart canals are shown in parentheses.

Designation

Southern subsystem:

S1)	Cerro Orejas MEC (modern Orejas)
S2)	Cerro Arena-Cerro Blanco MEC (Santo Domingo)
S3)	General de Moche (still in use)

Northern subsystem:

N1)	Punta Serrano MEC (modern Serrano)
N2)	Moro Viejo MEC (modern Moro)
N3)	Vinchansao MEC (modern Vinchansao)
N4)	Mochica Canal (modern Mochica-Valdivia)
N5)	La Cumbre-Vinchansao extension, or Rio Seco MEC (no modern counterpart)

Only the S3 (General) MEC retains its former status. All other pre-Hispanic canals have atrophied to varying degrees, and the intervalley N5 (Rio Seco) MEC is no longer operational. Table 1 summarizes the quantity of land embraced by the prehistoric system and contrasts this with the amount of land under cultivation in recent years. The latter figures are calculated from 1942 aerial photographs that were taken before the widespread introduction of well pumps, so that the table compares basically the same water management technology in ancient and modern settings.

Table 1 indicates that the contemporary canal system encompasses 128 km^2 or about 40 percent less land than the 211 km^2 embraced by ancient MECs. This disparity between ancient and modern irrigation is mitigated by a number of factors. First, the table contrasts two synchronic agricultural maximums -- the 1942 historic against the total prehistoric system. The former is associated with European-derived occupation less than 500 years old, whereas the latter is the product of three millenia of indigenous occupation. Second, not all ancient canals were necessarily in use or equally extensive at the same time. Third, one area into which prehistoric canals were pushed, Pampa Huanchaco, was not effectively farmed in its entirety. Fourth, the table includes areas formerly farmed by sunken fields that are now leveled and cultivated by canal irrigation. Fifth, ancient and modern agriculture depended upon different crops with somewhat different water and soil requirements. Sixth, and finally, although most well pumps are employed within the confines of modern canal-farmed areas as supplementary water sources, a minority are situated at the distal ends of the irrigation system and today there is somewhat more land under cultivation than in 1942.

In spite of such mitigating factors the contrast between ancient and modern times is dramatic and reflects a substantial degree of agrarian collapse. Identifying possible reasons for previous collapse is another contribution archaeolgists can offer specialists concerned with contemporary desert reclamation programs.

Early Agriculture

There is direct physical evidence of prehistoric reclamation activities in the Moche Valley in the form of abandoned canals, fields, and furrows. These remains date between about 500 B.C. and 1,500 A.D., and they survive because they were large undertakings that pushed beyond the confines of modern farming. However, these projects were based on long prior experience with agriculture that went on near the river in areas reworked by both later pre-Hispanic and modern farmers. Therefore, there are no physical survivals of the earliest agricultural systems, and the initial development of farming must be inferred from other lines of evidence including food remains and dessicated plants found in prehistoric refuse, and the spatial distributions of early settlements.

Economic evolution on the coast of Peru involved complex processes in which an early hunting and gathering way of life was replaced by a productive adaptation to fishing and littoral collecting, which in turn was displaced by irrigation agriculture. People first entered the coastal flats as hunter-gatherers exploiting flora and fauna of the river flood plains. Riparian resources were seasonal and not particularly abundant and this kept populations small, dispersed, and relatively mobile. People gradually expanded their resource base by utilizing marine foods and by tending plants, which by around 3,500 B.C. included gourds, squash, and legumes, presumably cultivated on self-watering lands.

Hunting-gathering declined around 3,000 B.C. as reliance on sea foods increased, and by 2,500 B.C. a widespread maritime adaptation crystalized along much of the Peruvian coast. It was supported by the richest marine biomass in the Western Hemisphere, if not all the world (Idyll 1973). The marine resources had a number of important characteristics. They were localized along the coastline, available year around, extremely abundant, and easily exploited. This economic base allowed people to reside in sedentary communities and it underwrote a marked population explosion, such that for every early hunter there were some 30 later fishermen (Moseley 1972). This increase resulted in coastal settlements of substantial size, often housing more than a thousand individuals. Demographically speaking, the "critical mass" requisites for complex society and the beginnings of civilization were reached. For unknown reasons the maritime populations began building platform mounds and other monumental architecture.

Corporate construction projects went on at many sites. The largest is the great masonry complex of El Paraiso, some 15 km. north of Lima. El Paraiso is a group of collapsed masonry buildings covering about 50 ha. that required on the order of 100,000 tons of quarried stone (Engel 1967). It was a prodigious undertaking in terms of any preliterate civilization and is particularly interesting because there was only a small residential population. Therefore, building El Paraiso apparently required the mobilization of labor forces on a regional basis drawing manpower from many different maritime communities.

A traditional supposition holds that only agriculturalists can engage in such undertakings, and that they do so at the instigation of priests controling the calendrics of planting and harvesting. Yet, analysis of food remains preserved by the arid environment indicates that the diet was based primarily on marine resources. Thus, the maritime economy supported not only large populations, but also the evolution or organizational principles that allowed a few individuals to direct the labor of many in public projects (Moseley 1975c).

Farming continued in use, and refuse analysis reveals minor quantities of legumes, squash, chili peppers, and wild fruits, as well as large amounts of cotton and gourd. The former crops served as condiments, not staples, and the latter were industrial cultigens. The emphasis on industrial crops is expectable. The sea could supply people's nutritional needs, but not their manufacturing wants. Cotton was needed for fishing lines, nets, and clothing, while gourds provided net floats and water containers (pottery not being in use at the time). Condiments and fruits introduced variation and nutritional balance in what was otherwise a rather monotonous marine diet. Because the demands on farming were not great, plant tending is inferred to have gone on in seasonally inundated flood plains or areas of high water table.

Most maritime sites are situated near the sea rather than near arable land. There are two such early settlements in the Moche Valley. Both sit atop wave-cut bluffs and the sea was presumably near the sites at the time of occupation. Padre Aban is located on the north side of the valley and the bluff it rests on is some 400 m. in from the modern beach. Alto Salaverry is on the south side of the valley and more than 1.5 km. inland (Figure 1).

The maritime societies may be said to have been preadapted to irrigation agriculture in three ways. First, the large labor forces that would be needed to build canal delivery systems already existed. Second, the organizational institutions necessary to instigate corporate construction projects had evolved. And, third, there were centuries of prior experience with plant tending. Of course one wonders why these societies should shift subsistence patterns. Intensive agriculture was underway elsewhere in the continent, but it was not supporting comparably large settlements, nor phenomena such as the building of vast public works. In fact, because of its support capacity the marine adaptation very likely inhibited much early reliance on coastal farming.

However, around 1,800 B.C. many of the maritime communities were abandoned, including most of the largest sites. The centers of monumental architecture shifted to inland settings, generally near valley necks and narrows. There was also a concomitant and marked increase in the plant foods being consumed, as well as the spread of new crops including tubers, peanuts, and maize. In the Moche drainage Padre Aban and Alto Salaverry were abandoned, and near the valley neck the Caballo Muerto mound complex developed into the dominant regional center. These events are presumed to mark the advent of irrigation agriculture and its rapid development due to cultural preadaptation. The factors triggering the shift are by no means clear. The abandonment of many, but not all, maritime settlements seems to indicate some sort of problem relating to the sea. Various explanations have been offered including over-exploitation of fish and shellfish, major niño cataclysms, and tectonic uplift which might have pushed fish into deeper, less accessable waters. Whatever the case, archeological evidence shows that marine resources continued to be important after the onset of intensive agriculture, as witnessed by the fishing village of Gramalote which was established several kilometers north of Padre Aban on the edge of the same bluff occupied by the abandoned community.

The location of Caballo Muerto near the valley neck may be explained in terms of the proposition that canal irrigation would most likely begin in interior settings. This argument rests on the presence of steeper gradients in the upper and middle valley areas, and lesser gradients near the coast. The hypothetical importance of this is threefold. First, in areas of higher gradients, leadoff canals required to channel water out of flood plains and onto the desert can be relatively short. Near the coast, where there is little slope, canal leadoffs must be substantially longer, requiring much larger labor investments. Second, once channeled onto arable land, it is easier to control water flow in somewhat inclined areas than on relatively horizontal surfaces. Third, areas close to the valley mouth have high water table conditions that often make irrigation impractical. It would be interesting to know exactly how much uplift, if any, the Moche drainage had experienced when the first canals were under construction. If there had been little or none, the water table may well have been substantially higher, with marshy conditions prevailing along lower sections of the river and behind the river mouth, which would also favor interior settings for canal intakes.

It is difficult to be very specific about the nature and location of the earliest canals. Small sites and settlements that would have grown up along their banks are either buried under meters of irrigation-deposited silt, or have been destroyed by later farming. Inferences must therefore be made from the location and distribution of the few large platform mounds that have survived. Platforms are present in the vicinity of the middle or initial sections of the S1 (Orejas), N1 (Serrano), N3 (Vinchansao), and N4 (Mochica) canals. The implication is that these canals, or more specifically their initial or middle sections, were in use before 500 B.C. and that the early canal courses were in about the same locations as their surviving counterparts. However, all of the early canals were later extended in length, and many if not all underwent elevational increase.

It is worth noting that Caballo Muerto, the dominant

early center in the Moche drainage, is adjacent to the N3 (Vinchansao) canal, which irrigated a majority of the best lands in the valley. Presumably, the association of this canal with this site is not fortuitous, but a reflection of the interrelationships reclamation shared with labor resources and socio-political power.

Later Prehistory

In the Moche Valley remnants of pre-Hispanic agricultural works survive from four periods of archaeological time. These temporal phases are called: Salinar (starting c. 500 B.C.); Gallinazo (c. 250 B.C.); Moche (c. 100 A.D.); and Chimu (c. 800 A.D.). Taken together they represent an uninterrupted cultural tradition. The phases are defined on the basis of changes in ceramics and art styles, and there is no evidence that the shifts from one phase to another coincided with major social or cultural change.

The agricultural works surviving in the archaeological record consist of sections of canals, fields, terraces, and sunken fields. There are more survivals from the Chimu and Moche phases than from the earlier phases. What survives comes from the outlying sections of the agricultural system, which is to say formerly irrigated areas beyond the margins of current cultivation. This distribution generates certain sampling problems, particularly for the early phases when canals were first being pushed well out into the desert. Today, the fringes of the irrigation system are high-risk areas due to insecure water supply, and farming there is done by small landholders and sharecroppers, while the central valley lands are corporately owned and worked. Thus, the archaeological data that survive are not necessarily indicative of what was going on in the center of the irrigation system.

The data are also biased by the asymmetrical profile of the valley. South of the river there are steep gradients and only moderate amounts of arable land. Here reclamation could proceed relatively rapidly by relying on canals of modest length. The opposite situation prevails north of the river. Shallow gradients meant that great expanses of arable land could only gradually be exploited through the slow construction of canals of large size and substantial length. Consequently, the earliest preserved water works are south of the river, while later, far more extensive remains are present in the north.

The earliest canal remnants identified at present are aspects of the S2 (Arena) MEC. To move water down-valley the S2 had to follow the contour around the point of Cerro

Arena, a ridge-like hill projecting out toward the river. Through time progressively more land was reclaimed by building new S2 channels that passed around the hill at successively higher altitudes following the principle of "elevation increase." On the summit and flanks of Cerro Arena there is a large Salinar phase settlement spread over several square kilometers. The settlement does not cross the S2 course or spill into the irrigated fields, and this is taken as an indication of contemporaneity between the site and one or more of the lower channels rounding the hill.

The S1 (Orejas) MEC underwent several stages of elevational increase and distal extension. In scattered areas below the highest channel there are Salinar artifact scatters. These may represent Salinar use of a lower, unpreserved S1 channel followed by subsequent site destruction when the S1 was raised and the area of former settlement brought under cultivation. Because the S3 (General) MEC is still operational there is no archaeological documentation of its development. There is also little archaeological knowledge of developments on the north side of the valley for either the Salinar or Gallinazo phases.

The Cerro Arena settlement was abandoned at an early date, and during the Gallinazo phase some people began settling around the nearby down-valley hill of Cerro Blanco. The S2 continued in use, and some of the middle level channels no doubt date to this phase. In fact, the shift in settlements was probably due to elevation of the canal, which opened new lands around Cerro Blanco. The largest Gallinazo settlement arose along the flanks of Cerro Orejas, and one or more of the lower S1 (Orejas) channels seem to be undertakings associated with this site. Along the margins of land presently irrigated by the S3 (General) MEC there is a string of Gallinazo cemeteries. The site distribution implies that this canal was established and probably nearing its maximum limits of expansion in Gallinazo times. Developments north of the river are not clear, largely because irrigation had, apparently, not been pushed beyond the margins of modern cultivation.

The Moche phase occupation can be divided into five subphases, during the first two of which the local population began to exert political influence over adjacent coastal drainages. By the third and fourth subphases the Moche Valley had become the seat of a polity controlling at least five other valleys. Situated at the base of Cerro Blanco, the site of Huaca del Sol and Huaca de la Luna (Shrine of the Sun and Shrine of the Moon) is all that is left of the imperial capital. The Huaca del Sol is a vast platform mound, the largest mud brick structure ever erected in South America. Huaca de la Luna is a somewhat smaller complex of three impressive adobe mounds. The flats separating the two huacas are now nearly level, but formerly they were occupied by a large array of public buildings and elaborate quarters. The two shrines and other monuments at the capital were built in segmented task units suggesting the employment of a mit'a-like labor tax as described above. The great size of the structures reflects the important political status of the site and the ability of the governing nobility to mobilize labor from many other valleys.

It was during Moche times that irrigation was pushed toward the maximum limits of the river's supply capacity. The highest and longest channels of the S1 (Orejas), S2 (Arena), N1 (Serrano) and N2 (Moro) were cut. By inference the S3 (General) probably reached full expansion. The N4 (Mochica) was pushed out to the middle of Pampa Esperanza and then carried seaward toward the coastal settlement of Huanchaco. Finally, in a massive undertaking the N3 (Vinchansao) was elevated and extended around Cerro Cabras to encompass the entirety of Pampa Esperanza. That the entire irrigation system underwent dramatic expansion, pushing far beyond the confines of modern cultivation, correlates directly with other evidence suggesting the high political status of the valley, the taxation system, and the tax investment strategy.

At the end of the fourth subphase of Moche occupation the economic and political fortunes of the valley changed. Huacas Sol and Luna were abandoned, and a new capital arose further north in the Lambayeque drainage. During the fifth subphase a large settlement was built at Galindo in the neck of the Moche Valley. Although Galindo spread over more than four km^2, there is relatively little monumental architecture at the site, and the few adobe platforms present are very small. There is no evidence that the existing canal system was expanded or that any significant additional reclamation took place. The marked decline in construction activity is not surprising. The valley was no longer in a position to draw on outside labor, but instead rendered taxes to the Lambayeque-based polity.

Chimu phase developments were in some ways a replay of the previous Moche phase occupation. Around 900 A.D. the local elite began re-exerting political influence in neighboring areas. Gradually hegemony was spread by force of arms or coercion and by about 1,450 A.D. the Moche Valley was the seat of an empire stretching from near the Ecuadorian border to central Peru just north of Lima. This was the largest state ever to arise on the Pacific coast and it controlled a majority of all the land reclaimed along the desert. Chan Chan was the imperial capital, situated near the coast north of the Moche River. It grew to be one of the very largest pre-Columbian cities in the New World, with massive monumental architecture spreading over six km^2 (Moseley 1975a).

Lowlands were brought under cultivation by constructing sunken fields. Much of this went on early in the history of Chan Chan. This was the first extensive local use of sunken fields, and it apparently relates to tectonic uplift, among other things. There is no evidence of human activity below the bluffs that were occupied by the earliest fishing communities until the Moche phase, when a settlement at Huanchaco included the construction of some houses along the base of the bluffs. By this time the valley had clearly experienced uplift, but how much is unestablished. The polity at Huacas Sol and Luna certainly had the labor resources to construct sunken fields, but for some reason they did not do so. It may be that they were unfamiliar with, or uninterested in the technique. Alternatively, the physical conditions necessary for sunken field farming may not have existed at the time. Whatever the case, the later Chimu did exploit substantial amounts of land by creating sunken fields. Along the edges of these fields small settlements and farmsteads often developed, such as Choroval on the south side of the valley mouth. The location of Choroval on the flats behind the modern beach contrasts sharply with the earlier maritime site of Alto Salaverry, situated much further inland atop an old wave-cut bluff. Here the archaeological evidence for uplift is straight forward, but when and at what rates this change took place is not entirely clear. Thus, the potential implications for changes in water table and opening of land to sunken field farming are still to be established.

Chimu canal construction activities focused on the northern flats of the valley. One early undertaking involved extending the N3 (Vinchansao) MEC out across Pampa Rio Seco to the edge of Pampa Huanchaco. This was later followed by the most ambitious irrigation program ever carried out in the region: construction of the La Cumbre intervalley canal and the N5 (Rio Seco) MEC which opened most of the Pampas Rio Seco and Huanchaco to cultivation. The La Cumbre canal is about 84 km. long, but the straight-line distance from channel intake to distal fields is only about 42 km. The difference is due to the canal contouring along very rugged Andean foothills before crossing an intervalley pass and descending into the Moche drainage above Chan Chan (Kus 1972). Prior to reaching the pass the canal had to cross extremely broken terrain with frequent bedrock declivities and slopes in excess of 60°. Although large earth-fill aqueducts were employed, steep rocky slopes were most often negotiated by running the channel along the top of vast artificial terraces upwards of 50 m. in height. Thus, building the canal was not simply a matter of cutting a channel. It entailed constructing vast terraces and aqueducts to carry a channel where none could be cut. In terms of labor investment and engineering the La Cumbre canal may prove to be the most prodigious hydraulic undertaking ever carried out in pre-Columbian America.

A 1,600 m. section of the La Cumbre canal near the intervalley divide has received detailed study, which indicates that Chimu engineers understood and manipulated the complex interrelationships of channel bed slope, side wall roughness, hydraulic radius, and sub- and super-critical chokes in a conscious manner that kept the canal's Froude number very close to unity. The Froude number of one represents maximum efficiency of water movement. What this study suggests is that a more efficient canal could not be designed today, and that native hydrologists employed concepts of fluid dynamics which were not known to western science until late in the 19th century (Ortloff 1977).

The La Cumbre canal linked up with the Moche Valley system at the top of Pampa Esperanza up-slope from Chan Chan, which occupies the base of the plain. At the linkup the Chicama waters were taken over by the N5 (Rio Seco) MEC and carried north and west for distribution across Pampas Esperanza, Rio Seco, and Huanchaco. The La Cumbre could have entered the Moche Valley by a much shorter route out of the intervalley pass if it had been intended to water only the latter two plains. Pampa Esperanza was already irrigated from the N3 (Vinchansao) MEC (Figure 1). Therefore, establishing the link-up above Chan Chan not only required a longer course and additional labor investments, but also placed the canal junction in an irrigated region.

At first glance this might appear a duplication and waste of effort. Actually, it was sophisticated water management strategy directed at water table manipulation. For potable water, Chan Chan depended upon 125 wells that drew on the aquifiers beneath Pampa Esperanza. The aquifiers were charged by irrigation agriculture on the north end of the plain between the canal linkup and the city. The Rio Moche is subject to marked seasonal fluctuations that would affect the farming cycle on the plain and generate fluctuations in the well levels at Chan Chan. The Rio Chicama is a far larger, perennial river. Bringing its waters into the head of Pampa Esperanza insured a relatively continuous irrigation regime and thus a constant supply of well water for the inhabitants of Chan Chan (Day 1974). The Chimu thereby combined irrigation with water table management in a creative way that goes beyond contemporary hydraulic strategies in the Andean desert.

About 70 years before Pizarro's conquistadores arrived, the Chimu and Inca empires came into conflict, and the former was conquered and incorporated by the latter. The Chimu dynasty continued to govern the valley, but was subservient to the new rulers who carved up the desert empire into separate fiefdoms allied to the Inca. Once Chan Chan was deprived of its former authority, no new desert areas were brought under cultivation. However, something significant happened to the city itself; people began farming within the formerly sacrosanct monument. Canals were cut through palace walls, and spacious open courts of the nobility were transformed into common fields. This was a strange, if not fitting end to what was once perhaps the most advanced center of hydraulic knowledge in the New World.

Agarian Collapse

In concluding, I wish to discuss agrarian collapse because it should be an important concern of people dealing with contemporary reclamation. Ongoing research will no doubt expand and refine the topic of collapse, but at this point some relevant observations can be put forward.

Uplift

Of the factors affecting coastal agriculture, tectonic uplift remains the most elusive. The configuration of the lower Moche Valley must have changed substantially between the time Alto Salaverry fishermen resided along the edge of a wide shallow bay stretching along the valley mouth, and when Chimu farmers at Choroval cultivated sunken fields in the dry flats of that bay.

Uplift probably correlates with the Rio Moche incising its course. Sometime after the Salinar phase the river downcut its channel by more than five meters. Minimally, this must have affected canal intakes and necessitated their reworking.

If uplift and downcutting were interrelated variables, then the entire irrigation system may have been affected in various ways. On the negative side the bed slopes of canals may have been altered. On the positive side, uplift and lowering of the river channel may have combined to depress the water table level around the valley mouth, thereby opening terrain to sunken field farming. I would further surmise that uplift has continued, allowing many sunken fields to be filled in, leveled, and irrigated for purposes of sugarcane cultivation in recent years. If this is correct, it is also worth noting that the now leveled sunken fields might more profitably have been kept in use, as many still are. The water required for their irrigation could then be used to bring other, dryer lands under cultivation.

Dune Formation

Dune formation is a less speculative matter. Today much of the south side of the valley is buried by deep deposits of unconsolidated aeolian sand. Saltating sand originates behind the beach in the vicinity of Puerto Salaverry and is moved inland by constant winds blowing off the ocean. Sheet sand and crescent-shaped dunes cross the valley mouth near Cerro Blanco, move over Cerro Arena, then lose momentum and blanket the landscape up to the base of Cerro Orejas. Dunes now cover formerly arable land at the lower end of the S1 (Orejas) MEC, the middle section of the S2 (Arena) MEC and adjacent areas, and many structures within the Salinar settlement at Cerro Arena. The settlement could not have been occupied nor the canals constructed if the sands had been present at the time. It can be argued with some certainty that the formation of active dunes was a relatively late prehistoric phenomenon and that it resulted in sand swamping the S2 MEC, causing its abandonment along with the distal end of the S1 MEC.

Dune formation and its associated agrarian collapse took place before Chimu times. There are no Chimu sites functionally associated with the canals or fields in the region. The Chimu built a long wall across the plain between Cerro Arena and Cerro Blanco, bisecting fields once watered by the S2 (Arena) MEC. The wall would have been functionally incompatible with the irrigation system, indicating that the system was no longer in use when it was built. Finally, adjacent to the river in front of Cerro Arena there are dunes with Chimu occupations on their tops.

At the Sol and Luna settlement there is evidence for the presence of saltating sands during the Moche phase occupation. Off the southeast corner of Sol residential architecture has been completely covered over with clean wind-deposited (aeolian) sand. The buildings are no earlier than the third Moche subphase, and intruded into the sand burying one room is a burial dating to the fourth subphase. Thus, sand buildup took place by or during the fourth subphase.

Furthermore, in the center of the plain between the two huacas there is a cemetery with Moche and Chimu burials in a low adobe platform. The cemetery is now level with the surrounding plain because the sides of the platform and areas between it and other buildings have been filled in by aeolian sand. At some point after the Moche burials the entire structure was soaked by flooding of the plain. The mud bricks of the platform partially dissolved and then solidified, encasing the burials in a solid matrix. The later Chimu graves were cut into the platform well after it had dried and were thus not encased in the matrix. It is unlikely that the top of the platform could have been soaked by flooding if sand had not filled in around it and brought the structure level with the surrounding plain. Deposits of aeolian sand were also encountered in a deep trench and two test pits opened in other areas of the plain. Thus, if the sand deposits at Sol and Luna are related to the dunes on the south side of the valley, as is highly probable, then dune formation either started during, or was underway by, the fourth subphase of the Moche occupation.

The impact of saltating sands on the occupation may well have contributed to the abandonment of Huacas Sol and Luna. The sand-covered architecture shows no reuse. The structures are very well preserved and show few signs of weathering or erosion prior to sand burial. This indicates that they were covered very shortly after they ceased to be used.

In summary, it is evident that dune formation began after Salinar reclamation activities on the south side of the valley. It is also clear that saltating sands had encroached upon the settlements near huacas Sol and Luna by or during the fourth Moche subphase, but this need not have been the direct cause of site abandonment. However, the dunes very likely caused the collapse of the S2 (Arena) MEC which brought water around Cerro Blanco to the foot of the Luna platform. The loss of this canal, and in particular the land it kept in vegetation down-valley and down wind from the site may well have intensified a deteriorating situation. Whatever the case, dune formation and agrarian collapse on the south side of the valley were probably complete by the fifth Moche subphase, and no doubt were significant in Galindo being founded on the north side of the valley. What exactly triggered dune formation is not clear. One might speculate that uplift exposed sand flats along the valley mouth, starting the dune formation. Yet, the emergence of such flats seems more of a precondition for dune development than it does a triggering mechanism.[1]

Flooding

A second case of agrarian collapse entailed massive destruction of the irrigation system by flooding. This flood was probably the same event that inundated the plain between Sol and Luna after it had been swept over by dunes. Although the entire valley was affected, the best evidence for flooding comes from Chimu canals on the north side of the valley. It is clear that canal destruction was not due to some extraordinary rise in the Rio Moche, because maximum elevation canals well beyond the river flood plain were washed out by waters descending from above the channels. Canals crossing normally dry washes well away from the river were uniformly washed out. Torrential rains are unequivocally implicated.

On the desert coast of Peru torrential rains are never localized events descending solely on one valley or region. Rains of consequence, such as those of 1925, are always associated with niños affecting vast stretches of the desert. Thus, the massive washout of the Chimu irrigation system can be attributed to niño-induced flooding. What distinguished this particular niño from others was the tremendous amount of flooding involved. Virtually every wash and small drainage carried water, and any place a canal crossed a topographic low it was washed out. Sheet wash spread across fields, eroding away furrows in some areas and silting them over in others, and it is doubtful that any aspect of the irrigation system escaped intact.

It is possible to be fairly specific about the intensity and amount of damage caused by this niño. It probably struck around 1,100 A.D., give or take a century. The canal system was later repaired. Rebuilt sections of channel are fairly easy to recognize and study. Where one is found stretching across a normally dry drainage it either remains intact to this day, or if washed out the damage is measurably less than the original washout that caused the section to be reconstructed in the first place. In other words, the combined erosional effect of all niño-induced floods and rains descending after the Chimu system was washed out and rebuilt has been less than the first great washout alone!

It is indeed fortunate that there has not been a niño of comparable magnitude during the last millenium because the impact of this phenomenon upon coastal civilization was truly staggering. All means and modes of agricultural production were destroyed, and the damage must have taken years to repair. Where canals were not washed out they were filled in by great quantities of alluvium, necessitating reconstruction of virtually the entire irrigation system. Most stored food must have been destroyed, because adobe buildings afford little protection against torrential rains. The Ñaymlap "legendary" account of the Lambayeque nobility mentions a cataclysmic flood of "30 days" duration followed by famine and pestilence during the Chimu phase (Kosok 1965). Nor does it seem entirely coincidental that the rulers of Chan Chan effected the conquest of Lambayeque shortly after the cataclysm. The Chimu had only to reconstruct two interrelated canals (the N3 [Vinchansao] which could supply the N5 [Rio Seco] independently of the La Cumbre canal) to regain economic footing, whereas rebuilding the vast Lambayeque irrigation complex entailed repairing a multitude of separate channels.

Somehow the Chan Chan dynasty recovered from this devastating niño and reconstructed nearly all the local irrigation system, with some losses on Pampa Huanchaco. These losses were, however, minor in comparison to the next disaster which led to the final collapse of the system.

Conquest

Agrarian collapse on the northwest side of the Moche Valley was late, massive, and swift. The N5 (Rio Seco) MEC dropped from use, as did most of the N3 (Vinchansao), and much of the N4. Pampas Huanchaco, Rio Seco, and the northern half of Esperanza were abandoned. There is no evidence of Spanish settlement on any of the lost lands, nor do any fields have straight furrows produced by animal-drawn plows. In fact, when the Spanish first entered the Moche Valley they made no mention of Chan Chan as a viable city, implying its prior abandonment.

It appears that collapse in the Moche Valley was intentionally instigated by the Incas as part of a conscious policy to undermine the economy of the Chan Chan dynasty following the conquest of the Chimu empire. There are no indications of an Inca presence on any of the lost lands, nor at Chan Chan. Yet, two things are clear. First, just before the Spanish arrival people began dismantling Chan Chan and farming in the ruins of the city. And, second, under Inca aegis a monumental compound based on Chan Chan prototypes was built in the Chicama Valley.

Given this situation I would argue the following. The Inca deliberately undermined the economic power and prestige of the Chimu, their largest and only effective rivals for Andean hegemony. In the Moche Valley this entailed cutting--literally or figuratively--the N3-N5 system and perhaps the N4, thus undermining the valley's agrarian base. If La

Cumbre canal was still functional, then it too was cut, returning the water to Chicama. The administrative functions formerly performed by Chan Chan were reduced and shifted to the new monumental compound in the Chicama. Finally, the Chimu nobility was either resettled within the Moche Valley, or they were forced to relocate when the land in front of Chan Chan reverted to desert and the city's wells ran dry.

The longer a complex irrigation system is out of use the more difficult it is to reactivate. That the Spanish made no attempt at reactivation is not surprising. They came to plunder, not to farm. By the time farming became a serious concern, the introduction of Old World diseases had decimated the native coastal population to a degree unmatched even by the great plagues of Europe. There simply wasn't the labor force needed for reclamation. After centuries of comparative calm and demographic regeneration, the labor force is now not only present but clamoring for land and food. But, the social system has changed, and priorities are not the same. A state-built industrial "park" sits on the formerly fertile fields of Pampa Esperanza. Chicken ranchers homestead the dry Rio Seco plains, depending on trucks to haul in potable water. A new sewage plant occupies the core of Pampa Huanchaco. People seem pleased at this "modernization", unaware that it proceeds at the cost of the best lands in the valley. Lost are the lands that fed past populations and could have fed future generations.

On the south side of the valley a small group of farmers attempt to scratch out a canal near the base of a large dune. The dune now straddles a dry channel constructed a few years ago by equally hopeful farmers. Far to the north bulldozers rumble in the Lambayeque Valley as foreign engineers direct them down the old Tayme canal. Yet, no one cares or wonders why either the Tayme or southern Moche lands fell into disuse in the first place.

Conclusions

I have discussed indigenous engineering and fluid dynamics in only a very general way because more is yet to be learned about these topics. Research in the Moche Valley demonstrates that they were highly complex and demonstrably productive. After all it was skill, not luck, that allowed early hydrologists to channel water over vast desert areas where none flows today. However, using modern know-how to reconstruct an extremely sophisticated ancient irrigation system is taxing, time consuming, and expensive. In some ways it can be compared to finding an alien space probe and from its wreckage attempting to reconstruct the technology and society that made it fly.

Therefore, I have pursued more of an historical narrative concerned with agrarian collapse. It is simply much easier to determine the reasons an agricultural system breaks down or a space probe crashes than it is to understand the technology and strategy that propelled either. If this paper shows skepticism about the empirical underpinnings of contemporary reclamation programs it is because these undertakings are seen through an archaeological eye. This is the eye of time. It does not measure the consequences of today's actions in years or decades, but in the centuries it will take for truly successful reclamation to feed future generations.

Acknowledgements

Information discussed was gathered with the assistance of National Science Foundation Grants GS-37765X and BNS 76-24538. I am indebted to the Programa Riego Antiguo staff. Of particular assistance were Thomas and Shelia Pozorski, James Kus, Lonnie Pippen, Fred Nials for geological data, Charles Ortloff for hydraulic data, Eric Deeds for data on flooding as well as his preparation of Figure 1 and Table 1, and Robert Feldman for editorial assistance.

Notes

[1]Nicholson (this volume) suggests that fluctuations in rainfall may have been an important factor.

References

Craig, Alan K. and Psuty, Norbert P.

1968 The Paracas Papers: Studies in Marine Desert Ecology (Reconnaissance Report), Florida Atlantic University, Department of Geography Occasional Publication No. 1.

1971 Paleoecology of Shell Mounds at Otuma, Peru. The Georgraphical Review 61 (1): 125-132.

Day, Kent C.

1974 Walk-In-Wells and Water Management at Chanchan, Peru. In The rise and fall of civilizations, modern archaeological approaches to ancient cultures, edited by C.C. Lamberg-Karlovsky and J.A. Sabloff,

Day, Kent C. (cont.)
pp. 182-190. Cummings Publishing Company, Menlo Park, California.

Engel, Frederic
1957 Early Sites on the Peruvian coast. Southwestern Journal of Anthropology 13:54-68.

1967 Le Complexe Preceramique d'El Paraiso (Perou). Journal de la Societe des Americanistes LV:43-96.

Idyll, C.P.
1973 The Anchovy Crisis. Scientific American 228(6):22-29.

Keatinge, Richard W.
1974 Chimu Rural Administrative Centers in the Moche Valley, Peru. World Archaeology 6(1):66-82.

Kosok, Paul
1965 Life, Land and Water in Ancient Peru. Long Island University Press, New York.

Kus, James S.
1972 Selected Aspects of Irrigated Agriculture in the Chimu Heartland, Peru. Unpublished PhD dissertation. Department of Geography, University of California, Los Angeles.

McGinnies, William G., Goldman, Bram J., and Paylore, Patricia, editors
1968 Deserts of the World, An Appraisal of Research into Their Physical and Biological Environments. The University of Arizona Press, Tucson.

Moseley, Michael E.
1972 Demography and Subsistence: An Example of Interaction from Prehistoric Peru. Southwestern Journal of Anthropology 28(1): 25-49.

1975a Chan Chan: Andean Alternative of the Preindustrial City. Science 187:219-225.

1975b Prehistoric Principles of Labor Organization in the Moche Valley, Peru. American Antiquity 40(2):191-196.

1975c The Maritime Foundation of Andean Civilization. Cummings Publishing Company, Menlo Park, California.

Ortloff, Charles
1977 Hydraulic Engineering Practice Among the Coastal Chimu, Part I: The La Cumbre (Intervalley) Canal System. Field Museum of Natural History, Chicago. Xeroxed.

Parsons, Jeffrey R. and Psuty, Norbert P.
1975 Sunken Fields and Prehispanic Subsistence of the Peruvian Coast. American Antiquity 40(3): 259-282.

Parsons, James J. and Denevan, William M.
1967 Pre-Columbian Ridged Fields. Scientific American 217(1): 92-100.

Richardson, James B., III
1973 The Prehistoric Sequence and the Pleistocene and Post-Pleistocene Climate of Northwest Peru. Variation in Anthropology, edited by Donald Lathrap and J. Douglas, pp. 199-211. Illinois Archaeological Survey.

1974 Holocene Beach Ridges Between the Chira River and Punta Parinas, Northwest Peru, and the Archaeological Sequence. Paper presented at the 1974 meetings of the Society for American Archaeology, Washington, D.C.

Robinson, David A.
1964 Peru in Four Dimensions. American Studies Press S.A., Lima, Peru.

Rowe, John H.
1948 The Kingdom of Chimor. Acta Americana 6(1-2):26-59.

West, Michael
1977 Alternatives to Canal Irrigation in a Coastal Peruvian Valley: Past and Present. Los Angeles County Museum of Natural History. Xeroxed.

Lees' paper contrasts local development schemes in nations with different socioeconomic and political systems. Although anthropology originally made its intellectual mark through the analysis of small-scale, traditional societies and cultures, its scholars are increasingly examining the features of industrial society itself. Nations and cities, as well as villages and tribes, have become our units of analysis. Today it is difficult to find peoples who are not in one way or another being brought into the sphere of the modern industrial world. The impact upon their way of life is colored by the nature of the philosophy and social organization of the larger political unit which dominates them. The anthropologist tries to lay bare the relations between the micro and macro levels. By this I mean two things; 1) that we are concerned to show how human behavior is conditioned by and reflective of institutionalized culture patterns, and 2) local communities are integral parts of a larger system, and must be understood as such. Lees' examples show that the enormous difficulties and frustrations often encountered at the local level may not be fully understandable nor correctable in terms of that level alone. The behavior of the farmer and the extension agent is conditioned by their entire milieu and the latter includes not only the nation-state itself, but the way in which that unit relates to the world.

2

Farmers and Technical Experts: Information Flow in Irrigated Agriculture

Susan H. Lees

Introduction

Much of the critical discussion focused upon contemporary human impact on arid environment ecosystems has been concerned with the negative consequences of poor planning, inappropriate technology, inadequate research, and prodigal use of hard-to-renew resources by market-oriented exploiters. It is doubtless true that such factors have contributed to the creation or extension of human-made deserts in some parts of the world, in the sense that ecological interference results in environments inhospitable to agriculture. But if planning were improved, technology better designed, research more extensive, and farmers more thrifty, would negative human impacts cease to occur?

It is indeed likely that many problems, particularly those associated with the construction and use of irrigation projects in arid lands, could be alleviated by such improvements. But no project is perfect, nor can any engineer or planner be expected to anticipate every problem that comes along. It would be valuable, then, to learn what happens _after_ irrigation projects have been implemented, in order to discover ways in which people who are affected respond to unanticipated problems. For if we agree that it may not be possible to plan for every contingency or to know in advance what all the relevant factors are, then clearly it would help us to know more about the consequences of unexpected developments in order better to deal with them as a class in the future.

It will be suggested in this paper that some, perhaps a great deal, of the damage that may occur to local environments after the project implementation stage, is a result of a kind of systemic insensitivity to what begin as small problems or accidents, but which grow to increasingly larger proportions if they are not arrested early. This systemic insensitivity itself is the result of poor communication links at critical junctures between farmers, the users of the new technology, technical experts such as extension service agents, and administrative intermediaries of various sorts. In order to illustrate what is meant by systemic insensitivity in concrete situations, I have drawn upon my fieldnotes from Mexico and Peru for examples which typify three different types of problem areas in information flow (1). This is not intended as a complete typology but a preliminary discussion of weakly integrated communication systems.

The second part of this paper is devoted to a discussion of a contrasting situation. Here, I use cases from Israel to illustrate the reverse: a very tight communications system and its apparent consequences (2). My major point is that while unanticipated problems may be inevitable in any irrigation system, waste and damage may be circumvented if the problems are checked in time. But appropriate timing may be possible only in certain contexts, as I shall show in the third part of this paper in which I discuss some differences between Israel and Latin America. This discussion is followed by suggestions on how social scientists might be useful at different points in the development effort.

The small-scale examples discussed here represent situations often found on a larger scale with far more serious consequences. The types of problems illustrated have been reported for such diverse areas as Bangladesh, the Philippines, and other parts of Latin America; they seem endemic in developing countries.

Case I: The Rusty Tractor Syndrome: Or, Who's in Charge of Repairs

Tractors are expensive, scarce, and very useful in the Valley of Oaxaca, Mexico; few individuals can afford to own them. I was surprised, then, to find one sitting idle in the middle of an empty lot in one village, rusting away, week after week. My inquiries about its ownership were met with unusual evasiveness and embarrassment. I finally learned that it was owned not by an individual, but by a small village cooperative, the *ejido*, which had received it through the agency of the ejidal bank. Members of the cooperative had shared in its use by turns, but one day it broke down, and the group could not arrive at a decision as to who was responsible for having it fixed. (There is a garage in the village; when privately owned tractors or trucks need repairs, they may be taken there or to a nearby town where several garages are available). As time went on, the co-operative became increasingly apprehensive lest the Bank discover that the tractor was out of use, though just what the consequences of discovery might be, were never spelled out to me. So that tractor remained where it was. This was not an unusual situation in this village. Where repairs and maintenance were a communal responsibility with no one officially delegated to the task, other facilities also often went neglected for considerable periods. The ejido's indecisiveness may simply have reflected a more pervasive pattern at the community level. The pattern was not general, however, for the region as a whole.

In this case, it seemed that little harm was done except for waste. But the problem of determining who is responsible for making repairs both at the farmer's level and at higher administrative levels, is a widespread obstacle to the effective use of new technology and resources. It can result in environmental damage when the unit in question is critical for deterring further decay in an already disturbed situation, or for salvaging a damaged system. For example, breakdown of a tubewell irrigation device may be insignificant when farmers plant traditional pre-irrigation crops, but disastrous when they plant the new "miracle" seeds of the Green Revolution. Both waste and damage have been reported as a consequence of the absence of an effective repair organization in a Bangladesh tubewell irrigation project.

Case II: The Virgin Canals Syndrome: Or, No More Help, Please

When the engineers drew up their plans to install a lifting device on a stream to bring more irrigation water to another village in Oaxaca, many of the farmers were somewhat skeptical about its prospects. Experience suggested that the chosen location was not the safest. But the experts had their way.

To do their part in this project, the villagers themselves were obliged to excavate new canals to utilize the water. They spent over six months in unpaid labor to complete the task. When the canals were finished and the lifting device installed they awaited their reward. But the heavy late spring rains had brought so much silt and debris down the stream that the lifting device was useless; the canals remained dry and never served any purpose at all.

But by the time of my visit, some months later, still no one had gone to the government office to inform the engineers of what had happened, nor had any engineer come to check the site. It was on record that the new device was now irrigating a specified number of additional hectares of land. The villagers seemed content to allow the deception to persist. Why? Not because there was no village official in charge, as in the last case. Rather, there was a widespread feeling that the reappearance of the engineers on the scene could only mean more trouble, more cost in effort and money. The villagers had already invested enough in a lost cause; they felt they had learned their lesson.

This syndrome resembles the last in that it involves waste, and may bring about some damage under special conditions. But it illustrates another point; that even where sufficient organization exists to carry out the task of communicating with technical experts, farmers may not do so anyway because they have learned to be distrustful of these experts. They have learned in particular that outside intervention will result in high costs but little progress. Communication between farmers and technical experts may, as a consequence, diminish even further.

Case III: The "My Superiors Have Not Yet Answered My Letter" Syndrome

The Peruvian agricultural ministry invited me to observe what was felt in Lima to be a fairly common problem: after farmers were provided with water by a new project, they refused to make use of it to irrigate their crops. However, one project area I visited in the southern Peruvian highlands provided me with a case study of a problem quite different from those perceived in Lima.

My official host, the on-location representative of the Ministry, was the technical expert assigned to oversee and supervise the irrigation system itself. This consisted of a large dam across the main river, a reservoir, and a canal system which led the stored water to the scattered hamlets of the region. The project had been funded by foreign aid and constructed by a foreign contracting company; it had been completed several years ago.

My host, the engineer, took me with him on several rounds of inspection, and together we also joined a team of social workers on their rounds to organize farmers into a cooperative. The latter trips were important, because one of the social workers could speak the local Indian language, which the technical expert could not. On occasion, the social worker would translate complaints and anxieties expressed by local people to the technical expert.

It did not, however, take words to convey one major problem certain farmers were experiencing, for it was evident to any on-the-spot observer. In several places, the canals had collapsed, with the result that water had leaked out to low-lying areas, creating useless marshes with a dissolved salt crust on top. These areas were once cultivated fields and pastures which now could not be employed at all. The farmers complained that they did not have the materials they needed to repair the canals nor the money to buy tools.

The technical expert was well aware of the difficulty. He explained (by way of his intermediary) that he also did not have the needed equipment, materials or funds, but that he had written some time ago to Lima for help. So far, he said, he had received no reply.

The situation of the technical expert in this case is not entirely irrelevant to the problem. Normally, directives come from above, from Lima, but information does not flow from the provinces back to the center. By reporting problems, rather than successes, the expert places a shadow over his own career; therefore, he fears reporting bad news. This is a hardship post; living conditions are the worst in Peru. The expert comes from hot country in the North; the topography, climate, and people are all unfamiliar to him. His wife is not happy. He would like to leave. He is a kind man and he cares about the people in his charge, but he does not have a great deal of self-confidence or assertiveness. If he did, he wouldn't be there.

This case represents the most serious type of communication-gap problem typical of technical development. In contrast to the others, it is not merely a case of a waste of time, money, effort and equipment, even though that may be the way it is seen at government headquarters. Rather, in this case, environmental damage has begun which, if not corrected soon, will spread and intensify to the point where it may be irreparable.

Here we might say that a set of social relationships and a political structure are as much at fault as inadequate planning and poor technical construction. The canals themselves might easily be repaired, but the administrative organization is such that those who are aware of the need for repair do not have the means at their disposal; those who have the means, lack the information needed to respond appropriately. Caught in the middle is a technical expert who has neither the power to use his technical expertise (for lack of funds) nor the strength to carry out a critical administrative function, informing his superiors of problems (for fear of losing his job).

Because of its own system of rewards, the administrative structure has nearly insured that the least effective technical experts will be found in the most remote and problem-ridden areas, on projects where planning and construction are most likely to be faulty, most in need of constant monitoring and adjustment. A technical expert who is well regarded by his superiors is likely to be rewarded with a more comfortable posting, closer to centers of wealth, power and progress. Working in the underdeveloped hinterlands is not seen to hold any rewards either as a route to promotion or as a virtue in itself.

To summarize the three types of problems illustrated here, the first was, in part, a problem of organization. That is, the obstacle to effective communication was the lack of an appointed individual or group in charge of reporting or correcting breakdowns. The second represented those problems arising from mistrust and misunderstanding. The engineers had ignored the advice and experience of the farmers, to the misfortune of the latter; they responded by withdrawing and avoided further communication. The third case illustrated problems arising from the bureaucratic context; farmers reported their problems, but their complaints never seemed to reach the individuals who had the power to respond to them.

Israeli Observations

In contrast to these situations, one can find development contexts which do not display such problems. While technical breakdowns and failures surely occur everywhere, in *these* situations a rapidly responding corps of technical experts is able to introduce considerable stability and avoid long-term deterioration and waste. This is not to say that as a general principle, speed of response and systemic stability need be associated, but that they *can* be in cases of technical development.

To illustrate the contrast, I will use incidents from rural Israel because I have had direct experience there, but it is likely that similar illustrations can be found elsewhere, in countries as different as the United States and perhaps, China. I will go on then to examine some factors that may be unique in the Israeli situation, although lessons learned there may suggest a variety of contexts in which a highly sensitive communication system can result in systemic stability. I should emphasize that I have been very selective in my choice of illustrations in order to make a point.

Case I: The Telephone Factor

The first factor I found in the Israeli context had to do with the technology of communication. The following case is illustrative. After a week of working in the kitchen peeling potatoes, I was finally assigned, as I had asked, to work as assistant to the irrigation manager of the northern border kibbutz I was visiting. This kibbutz had a unique system of irrigation, still in its experimental stage. The sprinklers in the orchards were run electronically. The irrigator simply switched pipes on and off from a central office on the kibbutz. Of course, more work than this was required; each day, he went to the fields to check each sprinkler to see that it was turning properly, repair broken mechanisms, and so forth. (This, by the way, had to be done by ear, since orchards do not permit much visual surveillance. Sprinklers "sing" their tunes; the experienced irrigator listens, and checks the ones he hears singing "off key.")

One morning I arrived to find there was something wrong; too many red lights were on at the control panel. The irrigation manager turned to the telephone and called the Haifa company which had installed the electronic system. Two hours later, the repairman arrived and fixed the broken device. No damage was done, no interruption of routine occurred.

I was even more impressed with the rapidity of reporting and response to technical problems when, one day, I was assigned to work in the fish ponds. Carefully managed fish ponds are found on this and other kibbutzim of the north. That day, the man in charge noticed that one pond was not draining properly. He went to his truck and radioed for help. Within the hour, the regional technical expert on fisheries had arrived with advice on how to proceed. The problem was solved; once again, with no delay, no interruption of routine.

Having telephones and automobile radios in this instance was absolutely critical for the maintenance of effective technology. The sensitive location of the kibbutz, right on the Lebanese border, might have contributed to the development of rapid communications procedures, though similar situations were found elsewhere far from the border.

Case II: The Muddy Feet Factor

While having the most modern technology of communication surely helps, it is not the only element responsible for the sensitivity and responsiveness of the Israeli system. Far more important is the amount of information which is already there, stored in the minds of the technical experts. If those who are capable of and responsible for correcting or adjusting a problem have first-hand familiarity with it, the corrective process will take place more rapidly and effectively. In the Israeli case, direct personal experience in the field, renewed frequently in the course of their regular duties, helps experts to respond quickly.

In the summer of 1973 I spent a few days with a technical expert from the office of the water commissioner. His job normally involved sitting at a desk in Tel Aviv evaluating plans and proposals for water use, but from time to time, he would drive out to farms for on-site observations. Two weeks before the time of my visit, a number of farms had had new pressure-gauge controls installed. Since then a few complaints had come in to the regional technical experts, who had passed on the information to the main irrigation office. A field visit was planned by the water commissioner's office representative, to be made in the company of regional officials. We drove to three sites that day, each one like the next. In each area, we first picked up a regional official and sometimes another expert, then drove out to pick up the irrigation manager of the kibbutz before inspecting the specific field, pipe or gauge in question.

The first visit was brief. It was to a muddy field. All of us removed our shoes, rolled up our pants and marched out between rows of cotton to test the pressure gauges ourselves. (I mention this in particular because I had heard of similar incidents before. In particular, one Israeli extension service agent who had been sent to Turkey on a foreign aid mission told me how he had shocked his Turkish counterpart by joining farmers in an irrigation ditch, barefoot). The pressure gauges in the first case were working properly; it turned out that the irrigation manager had misunderstood the new device. The second instance called for a quick adjustment, while the third involved replacement of some pipes. The central office technician called the factory, put the defective pipes in his car trunk, and made sure of the replacement personally.

Analysis

No technical installation is perfect; even with the finest planning and construction, unexpected occurrences necessitate adjustments. Major technical developments in agriculture anywhere, inevitably meet with problems. But what happens as a result of these problems is critical. It may well be that early detection and response can eliminate the negative environmental effects of many if not most unanticipated problems of new installations.

In the Israeli case, we can see how important a tight communications network might be in preventing negative effects of technical change in agriculture. One significant component of this network is a highly advanced communications technology, including telephone and radio systems not available in all developing areas. Perhaps the technical communications set-up should be considered an essential component of all new irrigation projects. Where not already available, it can be introduced as a new component like others more directly associated with the hydraulic installation itself.

But certainly, the telephones and the radios are only partially accountable for the effectiveness of the communications network. In some ways they play the least important part. Several other factors are more critical, including the fact that there are a large number of technical experts, proportionately, in relation to the number of farmers. These experts are available in the extension service, and in private concerns. They are not only numerous, but highly accessible, and can be on the spot quickly in a country as small as Israel. But, in addition to this, a number of other factors should be considered.

1. Israelis who have gone through both the Israeli school system and the army, have first-hand familiarity with much of the country. As part of their education and training as citizens, they travel to all parts of the country. As members of youth groups or high school classes they visit kibbutzim and work several weeks of the year in agriculture. They learn first-hand what it means to harvest, plow, weed and irrigate. This shared experience facilitates communication between rural and urban Israelis in a way that is unusual in other developing countries with the possible exceptions of China and Cuba.

As a result, one is unlikely to find a technical expert in agriculture who does not know something about the region in which he or she is working. Their decisions are made, whether in the fields or in central offices, on the basis of direct experience.

2. The rural population of Israel is generally described as having three major components: Palestinian Arabs, Jews of Occidental (Ashkenazi) origins and Jews of Oriental (Sephardic) origins. Most of the latter come from North

Africa and the Middle East. Agricultural experts work with all three, and find themselves in quite different social and technical contexts with each of the three groups. Most technical experts in agriculture are of Occidental origin. They, most other officials, and the majority of the farmers themselves tend to be of the same ethnic background, hence there are few problems of misunderstanding on that count.

In cases where technical experts are dealing with farmers of Oriental origins, ethnic differences apparently do (or at least did, in the recent past) impede communication. However, I have too little information on such cases to elaborate more fully. Although Hebrew is a second language for many Israelis, language barriers seemed to be minimal or absent in nearly all cases. Those experts I knew who worked with Arabs were all quite fluent in Arabic.

While Israel shares with many other developing countries the potential for problems of communication due to ethnic and linguistic diversity, these turn out to be far less of an obstacle than elsewhere, thanks in part to concerted efforts by much of the population who, as immigrants or recent descendants of immigrants, hope to minimize past differences and reinforce a sense of common interest and shared experience.

3. The most significant contrast, in my view, with so many developing countries, relates to class divisions. Both the ideology and the practice of socialism in Israel have tended to minimize the development of a class structure. While differences of income, education, and standard of living exist, these differences do not reflect a rural-urban dichotomy. Among the Israelis, there seems to be no clear "peasant class" or "rural proletariat" as elsewhere (though there may be both within the Arab population). Consequently, impediments to communication resulting from class divisions appear to be absent in the Israeli situation. It may be, again, that China, Cuba, and some other developing nations share this advantage as well.

In Israel, this factor is associated with the pervasive Zionist ideology of "return to the land" and "turn the desert green" which began several generations back. As a consequence of this ideology, farmers are regarded with respect, and officials and experts are far from reluctant to dirty their hands (and feet) in order to join them in their fields to understand their problems. The reverse situation obtains in most developing countries where the stigma of manual labor, particularly that of peasants, discourages mutual participation in agriculture. Thus, while the Israeli farmers and their experts communicated directly and easily, their counterparts elsewhere tended to regard one another with mutual distrust.

4. A final point, by no means the least important, is this: The socialist political and economic context of agriculture promotes direct participation of Israeli farmers in decisions affecting their lives. The agricultural sector tends, in fact, to be represented in government beyond its proportion in numbers of economic contribution. This helps to make centralized policy and administration sensitive to local and sectorial needs. Consequently, measures taken in agricultural development tend to be more responsive, hence more appropriate in most cases, to local ecological, economic and social conditions. The reverse is true in other developing countries, again with certain obvious exceptions.

Conclusions

Most developing countries differ greatly from Israel in their history, demographic composition, size, literacy level, class structure, ideology, economic status, military situation, and many other relevant factors. For this reason, it is clear that comparisons will always be conditional, and that lessons learned in Israel are not directly applicable elsewhere. But certain specific similarities between Israel and such vastly different countries as China, the United States, Cuba, Tanzania and others, suggest some general principles which might be useful in application to any situation of agricultural development.

One conclusion we might draw from the comparisons made in this paper is that an important component of development planning, too often neglected in project design and implementation, is the establishment of an early-warning quick response system. No technical installation in agriculture can be expected to run flawlessly. Some problems can be anticipated, such as, perhaps, drainage problems in irrigation. These problems might be detected and controlled through regular and frequent on-site checks by appointed experts and managers. But more serious and destructive are the unanticipated problems whose effects may well be cumulative. Detection of and response to these depend very much upon human factors. Farmers and on-location managers and experts are likely to sense the problems fairly quickly, but whether these are reported depends on what they anticipate as a response from others. And whether correction of the problem takes place, and how soon, depends also on a host of other factors.

Where class, language and ethnic barriers pose obstacles to communication and mutual trust, building sensitivity and responsiveness into the design of a project requires special foresight and ingenuity. It may be that the easiest place to begin is in preparing the supra-local network to respond to problems, such as insuring that spare parts are always available and rapidly delivered when requested. It has been suggested that in some countries the private sector might handle such services more effectively than the government. Decentralization of authority may also enhance sensitivity and responsiveness of development agency branches to local problems. This does not solve all problems due to planning and technical errors but it facilitates an ongoing adjustment to changing situations, which may help in avoiding waste of resources and efforts.

In future studies and evaluation of development projects, it would be useful to examine their design and operation with respect to degree of sensitivity to local disturbances, points of least sensitivity, and the time necessary for response to occur. One aspect worth examining is the technology of communication - telephones, radios, and road systems. The organization of communication, particularly the amount of paperwork that goes into reporting and responding to problems should also be investigated. Perhaps even more important in the communications network is the familiarity of technical experts with the region about which they make decisions. How much territory do they have to cover, how diverse is it, how long have they had direct personal experience there? Are there opportunities for farmers and technical experts to become acquainted personally?

Several papers in this symposium have recommended closer working ties between social scientists and other specialists in development planning and implementation. In the areas of communication, information flow and responsiveness to problems, social scientists could make extremely valuable contributions by pinpointing a variety of problem areas or suggesting ways of avoiding them. An anthropologist might, for example, provide information about the most effective communication channels already existing within a community, so that project designers might incorporate these as a structural part of the projects' organization. Social scientists could also make on-going studies of projects after implementation has begun. By detecting and reporting the changes which take place in social organization and behavior, and by suggesting project adjustments to these changes, they might make an enormous contribution.

This paper has addressed only one aspect of what many believe is a worsening situation in arid lands agriculture. It discussed some situations of waste, failure and environmental destruction resulting from poor information flow from farmers to technical experts and officials and vice versa. While most development specialists will concede that there is is room for considerable improvement in planning and technical design of arid lands projects, this paper suggests that even the best designed projects are likely to encounter some problems which call for readjustment to changing circumstances. The Latin American cases described here were intended to illustrate why the needed readjustments may not be made in certain development contexts. The Israeli case was used to illustrate not only the point that rapid communications can effectively reduce problems related to readjustent, but also the point that a rapid communication network develops more easily in certain social contexts than in others. An attempt was made to show that such contexts are characterized by a certain ideological, political,social and economic climate which is not present in many of the developing countries. It was not suggested that such a climate is absolutely necessary to bring about development, or to avoid waste and environmental damage, nor that it is any guarantee of agricultural success. Rather, it was suggested that a transfer of technology from countries like Israel, United States, and perhaps Cuba and China, to countries such as Peru, Mexico, or Brazil, for example, may well run into a variety of problems due to differences in information flow and administrative responsiveness. These in turn reflect profound differences in social, political, and economic structure. Avoiding or diminishing such problems will require planning, study and investment of effort of a sort not yet attempted on any wide scale in these countries.

Notes

1. The fieldwork upon which the Mexican illustrations are based was carried out during the years 1967-70 in association with a multi-disciplinary project, "The Prehistoric Cultural Ecology of the Valley of Oaxaca," funded by the National Science Foundation. My own research support came from NSF Grant G-2121, a graduate fellowship from the National Institute of Mental Health, and a Ford Foundation travel grant.

The fieldwork which produced the Peruvian illustration was carried out during the summer of 1974 in association with a two-year project organized by TAHAL, an Israeli consulting firm, funded by the Interamerican Development Bank.

2. The Israeli illustrations are based upon fieldwork carried out primarily during the summer of 1973, funded by a grant from the City University of New York Faculty Research Award Program.

Bedoian's work is a kind of ethnographic "first". Part of a multidisciplinary team sent to examine drought conditions in the Sahel, he collected the kind of specific and detailed information on traditional agricultural and herding practices necessary in order to assess their actual impact on the ecosystem. Further, he was able to provide data for comparing these more traditional practices both with those resulting from increasing population pressure in the area and with those recommended by modern range managers. The results are fascinating and will surprise some. They certainly deserve to be taken seriously by development-minded readers.

Human Use of the Pre-Saharan Ecosystem and Its Impact on Desertization

3

William H. Bedoian

Introduction

The productivity of the desert environment on the Saharan fringe varies greatly from year to year. Based on the experience of the Sahel drought and subsequent ecological collapse, the Saharan fringe seems to be easily upset by unwise human use under certain conditions. Specifically, when grazing and agricultural exploitation occur during periods of low rainfall, vegetation and soil stability are disrupted over vast areas. Without the vegetation to hold the soil, it is blown off the ground surface and deposited in dunes. In spite of such disastrous results, human populations continue to utilize the remaining usable land in the Saharan fringe in ways which lead to desertization. Desertization is the "spread of desert-like conditions in arid or semi-arid areas due to man's influence or to climatic change" (Rapp, 1974).

In 1972, the Tunisian PreSaharan Project, under the auspices of the U. S. International Biological Programme, initiated ecological research on the structure and function of the PreSaharan ecosystem on the Saharan fringe in Southern Tunisia. By including the human population in the

---

Acknowledgement: This research originated as a doctoral dissertation project in anthropology funded by a PHS-NIGMS Predoctoral Traineeship in Human Biology (training grant 5 T01 GMO-1748) administered by Paul T. Baker. Additional research funds were provided by a Smithsonian foreign currency grant (SFG-2-3492) to the Tunisian PreSaharan Project, Frederic H. Wagner, principal investigator. This paper benefited by suggestions from S. T. McGarvey and T. Gage which are gratefully acknowledged.

ecosystem study, it was hoped that the reasons behind current land use strategies could be determined. In addition, the extent of the human contribution to desertization would be assessed. In order to accomplish these objectives, a study of energy flow through the human population was undertaken in 1973-75. From the results of the ecosystem study, it was anticipated that less destructive land use alternatives could be proposed which would be compatible with the subsistence needs of the local inhabitants. The adoption of such alternatives would reduce or halt the processes associated with desertization.

The Energy Flow Approach to the Study of Human Ecology

In one sense ecology can be thought of as the study of "any property that has a measurable direct influence on the fecundity, longevity, speed of development, or spatial position of an organism" (Maelzor, 1965). Environmental energy flow influences the nutritional status and spatial position of organisms and can be considered as an ecological factor which may elicit a population response. The biological and behavioral attributes of a population can be analyzed as responses to ecological factors. Such responses have a bearing on the survival, reproduction, development, or spatial position of individuals in the population (Little and Morren, 1976).

Some components of the environment can be considered limiting factors on the amount of energy available to and trapped by ecological communities. Rainfall and temperature are such factors in the Tunisian desert, where rainfall and vegetal growth occur during the winter. Energy flow itself can be a limiting factor influencing the biological and behavioral responses of a population. An interruption in energy flow threatens the survival of man and other organisms in the ecosystem. If energy flow is near the minimal level needed to support a human population then one would expect certain behavioral and biological responses which would maximize the production of energy or minimize the utilization of energy.

The study of environmental energy flow is a method which can be used to evaluate the human exploitation of an environment. It is an especially useful approach for studying a population which derives most of its subsistence needs from the ecosystem in which it lives. One focus of this approach is the family economy. Observations are made on the production of crops and animals, on dietary consumption of the food which is produced, and on expenditure of human energy in work tasks associated with gaining a livelihood. Attention is also focused on the extraction from the environment of products which are not part of the human diet. The harvesting of firewood, of plants for animal feed, of wood or branches for building material, and of other non-food items is studied in order to assess the total use of an ecosystem by the human population within it. The study of energy flow is a refinement of previous anthropological approaches dealing with the human economy because with it the investigator can compare the energy value of wage and subsistence labor with the energy value of crops and animal products and with the energy requirements of family nutrition. The energy value of crops produced can be evaluated as a function of energy spent in work efforts. This offers insights not provided by considerations of man-hours in subsistence pursuits, or by estimation of cash value of subsistence labor, or by evaluations of cash flow.

In such studies the energy produced or extracted from the local ecosystem is traced through the various pathways of deployment by the family. Nutritional needs of the family and its domestic animals are primary concerns in this deployment. The storage of any surplus energy is seen as a strategy to deal with annual variability in energy flow or economic production. The import and export of energy to and from the local ecosystem is also studied, as is the amount of wage labor performed by members of the population. A knowledge of the various transfers of energy enables one to assess the degree of dependence of the human population on the local environment for meeting human needs and wants.

The Value of Energy Flow Studies for Planners of Economic Policy in Developing Arid Regions

If a population cannot derive all of its needs from subsistence pursuits in the local area, then it must depend on the import of food products to meet its needs. Human populations inhabiting arid lands are often characterized by some degree of reliance on domestic animal populations to furnish resources for the human economy. In such economies one often finds that semipastoralists convert animal resources into other goods (manufactured products, agricultural products, cash reserves) via some type of exchange or market system.

Energy flow research provides a knowledge of the subsistence alternatives available in an area and provides an understanding of the ways by which the inhabitants decide between alternatives. Often these alternatives are based on the ability to convert from one resource to another via a market system of some sort. One of the basic strengths of the energy flow approach is its ability to establish conversions or equivalencies between human labor (both subsistence effort and wage labor), money resources, agricultural and pastoral goods produced, and purchased food items. With this type of information it is possible to determine the relative costs and benefits of local food production versus food importation, of wage labor income versus income from the sale of animal and agricultural products, or of a reliance on one or another crop complex.

This approach is one of the few which enables the policy planner to treat unpaid agricultural and pastoral labor as a family resource. By determining the energy cost of various subsistence activities one gains a refinement which is not usually available from studies in economic anthropology. In calorie terms, man-hours are not equivalent to woman-hours or to child-hours. Each calorie of energy spent in subsistence tasks brings a certain return of energy in the crop harvested.

By simultaneously studying the deployment or investment of labor and material resources, one can assess the feasibility of potential economic development proposals which involve the use of land or labor in an arid region. If the population has a very limited margin of surplus resources, then it will be unable to survive a development program requiring that basic resources be set aside and invested rather than be used for a more short term direct gain to the family.

Economic Production In Southern Tunisia

Two aspects of material production must be considered when evaluating the effectiveness of population responses to the ecological constraints on subsistence production in Southern Tunisia. The first is the average amount of goods produced by the family over a period of many years. A legitimate question is whether this average production is enough to sustain the needs of the population without additional input from other ecosystems. The second aspect of land use is the extreme variation in annual rainfall, and hence in economic production from year to year in Southern Tunisia. The question then becomes whether agricultural and pastoral production is sufficient in excellent years to provide a surplus which can be stored and used in years of little or no rainfall. This question suggests very different expectations concerning the appropriate type of behavioral response on the part of the local inhabitants. Both the annual variation and the overall average production will be dealt with in this paper in an attempt to describe the strategies enabling the population to persist over time in this area of Southern Tunisia.

In Southern Tunisia one encounters a sparsely settled population of relatively sedentary Arab families who engage in agricultural and pastoral pursuits to gain their livelihood. Formerly the population was more nomadic and pastoral pursuits dominated the family economy. The economy presently includes widespread use of tractor services and human labor to produce barley, wheat and olives, and involves reliance on cash from the sale of male animals from family herds. In addition to the energy extracted from the local ecosystem, family resources are supplemented with cash earned by adult males performing wage labor in cities and towns in Southern Tunisia and Libya. Males leave the ecosystem to earn a wage income during the portion of the year when agricultural tasks are at a minimum.

The location of the study area of the Tunisian Pre-Saharan project and of the human energy flow project can be seen in Figures 1 and 2. The sampling zone of the human energy flow project was purposely located around the study site of the Tunsian PreSaharan project. Rainfall is continuously monitored on the study site and values reported here are derived from those data. A questionnaire concerning agricultural production between 1972 and 1975 was administered to 24 families whose home base falls within the sampling zone. Information was obtained on the size of barley, wheat, and olive harvests; the number of olive trees owned; the number of animals sold and the resulting cash obtained; the amount of human and animal food purchased; the amount of income from wage labor; the amount of cash spent on tractor services; the deployment of agricultural produce to the family and herd. Average harvest sizes were converted to energy production estimates by using the calorie value of barley, wheat, and olive oil reported by Leung (1968).

Observations on human energy expenditure in subsistence tasks were based on working time per unit of crop produced. For instance, the harvesting time for barley and wheat was recorded as minutes of work per kilogram of grain harvested. A sample of working individuals was observed for each task. The average harvesting time per kilogram of grain was then multiplied by the total harvest size in order to estimate the total harvesting time needed for the whole grain crop.

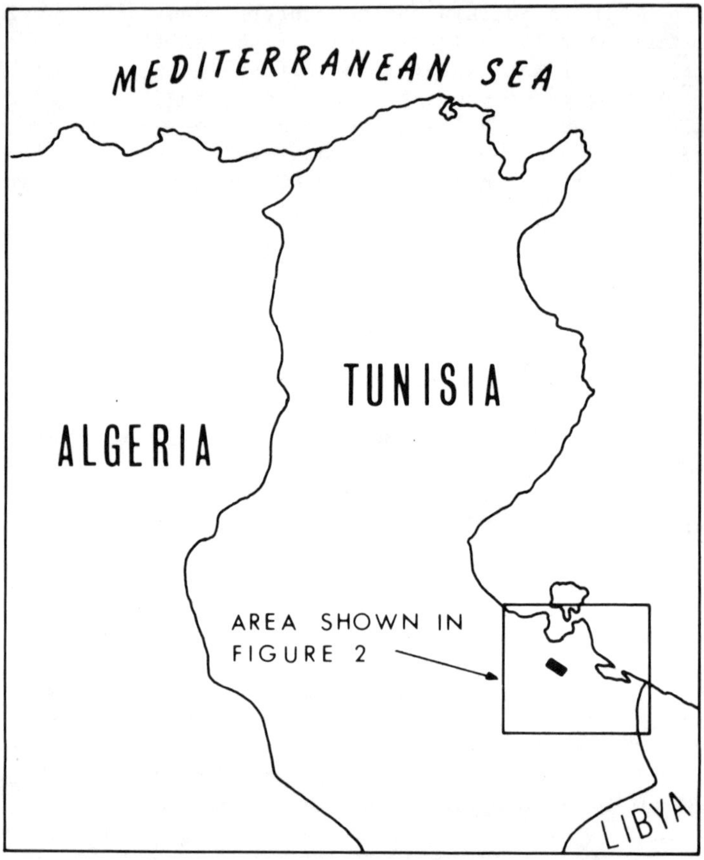

FIGURE 1. LOCATION OF STUDY AREA IN TUNISIA

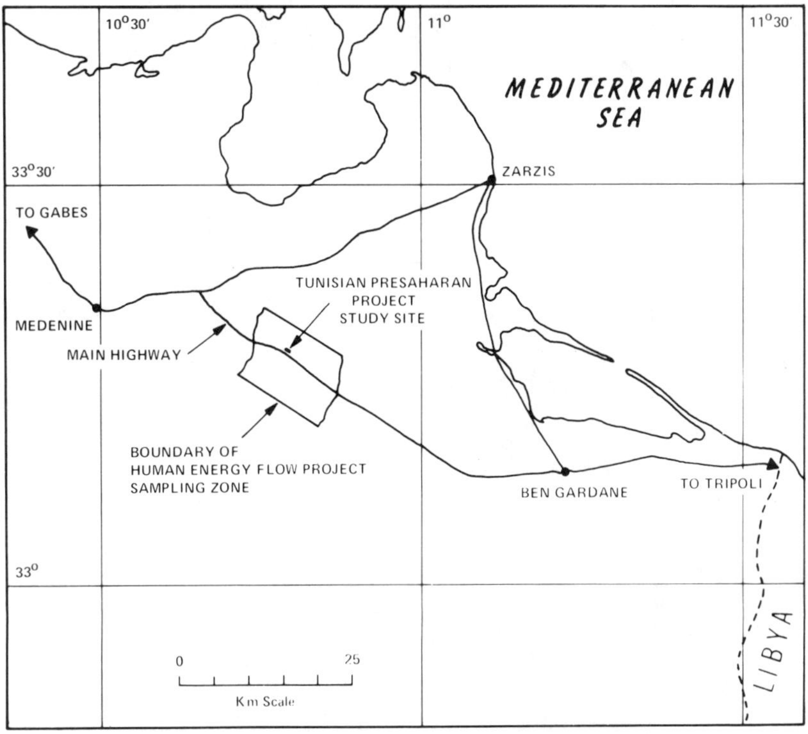

FIGURE 2. LOCATION OF STUDY AREA IN RELATION TO THE STUDY SITE OF THE TUNISIAN PRESAHARAN PROJECT

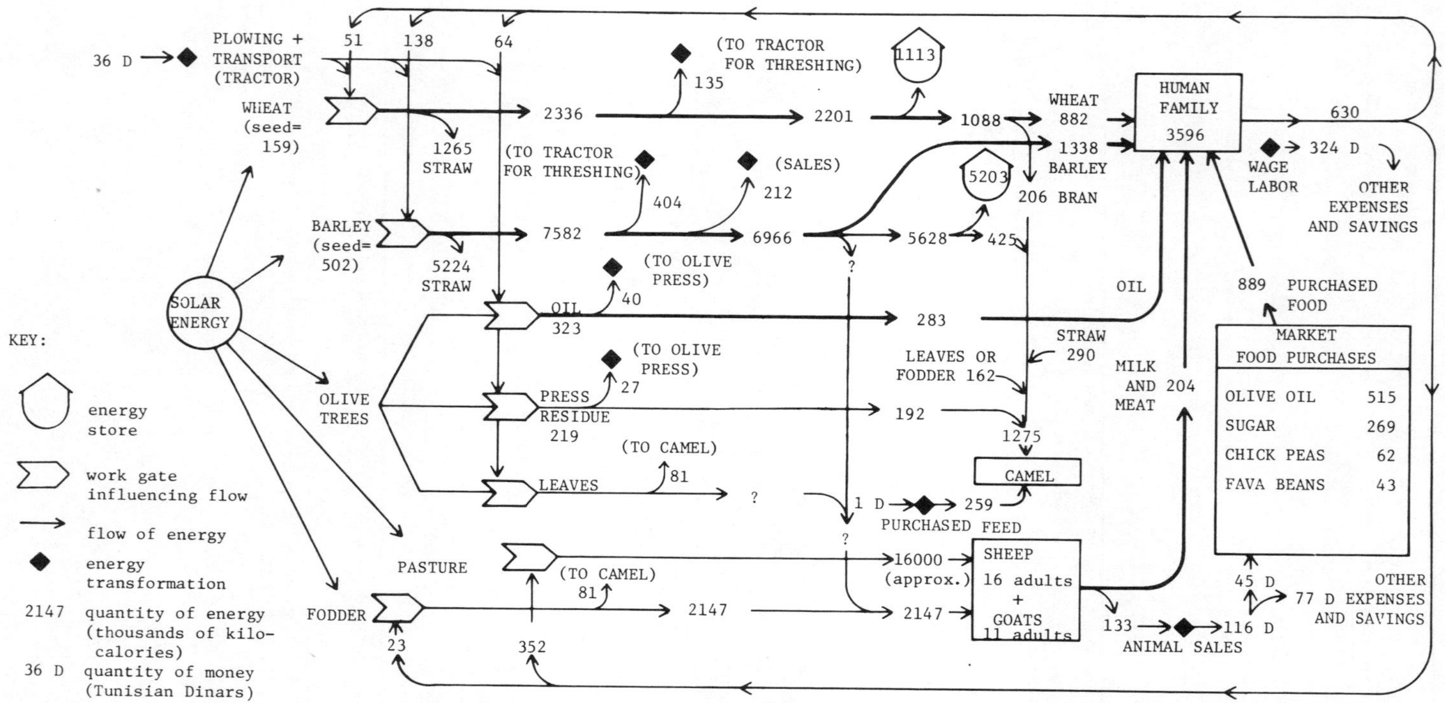

FIGURE 3. ENERGY FLOW (DURING 1974-75) THROUGH THE HUMAN POPULATION IN THE PRESAHARAN ECOSYSTEM.

The total working time for each subsistence activity was converted to energy expenditure as follows. The heart rate of working individuals was obtained during the performance of subsistence tasks. The average heart rate for each task was converted to an energy expenditure value by measuring heart rate and oxygen uptake for a sample of males and females during an exercise test. The results provide a regression of caloric expenditure on heart rate which can be used to predict a specific energy expenditure rate from a given heart rate. Thus the field measurements and the questionnaire data were combined to provide estimates of average energy expenditure and production associated with each crop produced during the three year study period. These averages form the bases of Figures 3, 4, 5, and 7.

The sample averages derived from the questionnaires are used to characterize the average family. Such a family consists of two adults and four children. It uses 69 hectares of land near its home base in its economic endeavors. Of this, 14 hectares are owned and the rest are borrowed from absentee land owners who may also be relatives. In addition, the family owns 16 hectares of land in small parcels scattered throughout the PreSaharan zone in Southern Tunisia. This land is used in years when rainfall on the home base has been inadequate for planting grain.

Economic Production During a Year of Abundant Rainfall (1974-1975)

During the agricultural year 1974 - 1975 (fall 1974 through summer 1975) 180mm of rainfall was measured on the study site. Figure 3 is a schematic chart of energy flow through the average human family and its herd for the 1974 - 75 agricultural cycle. In this diagram the flow of energy produced is read from left to right, and the paths and arrows indicate the deployment of energy produced. The energy values of all work activities needed to produce a particular crop have been combined into a single figure. This amount of energy is indicated by a vertical arrow leading to the work gate. The work gate symbolizes the work energy input needed to produce the crop. The producer organisms (plants) are grouped on the left, and the consumer organisms (grazers and humans) are grouped to the right. On the extreme right the exchange system is represented. The only purchased foods reported are those which make a large contribution to the caloric intake of the family. Items not reported are purchased in very small quantities each year or have a low calorie content. The diamond shaped symbols represent conversion of energy from one form into another. Much of the converted energy leaves the ecosystem as payment for mechanized services rendered. Therefore there are no arrows appearing beyond the conversion symbol. Cash derived from wage labor comes from outside the ecosystem as well, so there is no arrow connecting it to the local energy flow system. Cash income from wage labor and animal sales is used for the various expenses indicated in Figure 3. There are no connecting arrows between the sources of cash and various points of expenditure in the system. These have been omitted to keep the diagram as simple as possible. There are no connecting links between straw produced and straw consumed by the camel. Nor are there links illustrating the channeling of olive leaves and fodder to the camel. These omissions also make the diagram more readable. Food energy consumed by the family has been indicated by heavy lines. Most of the energy consumed by the family is spent in basal metabolism or in social activities. This is not represented in the diagram. Only the energy associated with subsistence work is shown. It is channeled back into the productive system in the various subsistence tasks.

Grain production. From Figure 3 several inferences can be made concerning the adequacy of this particular year of agricultural and pastoral production for meeting family needs. The amount of grain produced was greater than the amount used in the family diet. Because of the abundant rainfall, the herd only needed minimal food supplementation during late summer and early fall. It is likely that none of the animal diet supplementation was purchased. A considerable amount of grain was put into storage to be used during a year of poor rainfall. The family purchased both wheat and barley to obtain extra seed. This purchase was made because not enough seed was available from the family's grain reserves to take advantage of the abundant rainfall. A considerable amount of money was spent to engage a tractor to assist in this planting.

Olive production. The production of olive oil was clearly below the average family's annual need. Olive yields respond to rainfall occurring one year prior to the winter harvest, so that the yields of December 1974 - January 1975 were a reflection of poor rainfall the year before. A factor which tends to reduce olive crop yields is the presence during the spring of severe, hot, southwesterly winds (sirocco) originating from the Sahara. These winds blow most of the olive flowers off the trees before they can set fruit. This, along with the below average rainfall, probably was responsible for the mediocre olive crop in 1974-75. Poor olive production resulted in the purchase of substantial amounts of olive oil for family consumption during 1974 - 75.

Fodder harvest. Annual herbs and perennials were harvested in spring 1975 to be stored as fodder and used as a dietary supplement for the herd during the late summer months. The amount collected in 1974 -75 was about double that collected in each of the previous years. Because of abundant forage plants on the range land, it is possible that much of the fodder was not needed in 1974 - 75 and went into storage to be used during a dry year.

Cash income and expenses. Animal sales and wage labor produced a total cash income of 440 Dinars in 1974 - 75. Considering routine costs of purchased food, tractor service fees, and other incidentals, only about 90 Dinars of expenditures can be accounted for on the survey forms. This leaves an apparently large cash surplus during that year. However, since agricultural production was high in that year, the inhabitants arranged marriages, circumcisions, house building, and other social occasions which involved considerable expenses and food. Questions about such expenses were not included in the economic survey. In addition, families periodically use extra cash to buy up small parcels of available land adjacent to their home base land holdings. A few families have considered purchasing a share in a tractor, which is a major expense involving hundreds of Dinars. Although one might expect a family to attempt to build a cash reserve in a good year, the major unaccounted expenses discussed above probably reduce that reserve substantially below what might be estimated.

In overall perspective the 1974 - 75 agricultural cycle was the closest to a state of complete population support by the local ecosystem of any of the three years to be described. Of course, certain food items must always be purchased from outside of this ecosystem regardless of the productivity in any given year. But in the year just described, animal production was such that enough energy could be channeled into the exchange system to obtain those food items which could not be produced locally. Grain production was sufficient to create a large reserve for use in a drier year.

Economic Production During a Year of Near Average Rainfall (1973-1974)

The beginning of the agricultural cycle in the fall of 1973 was delayed because of below normal autumn rains. A sufficient rainfall occurred in mid-December but because of

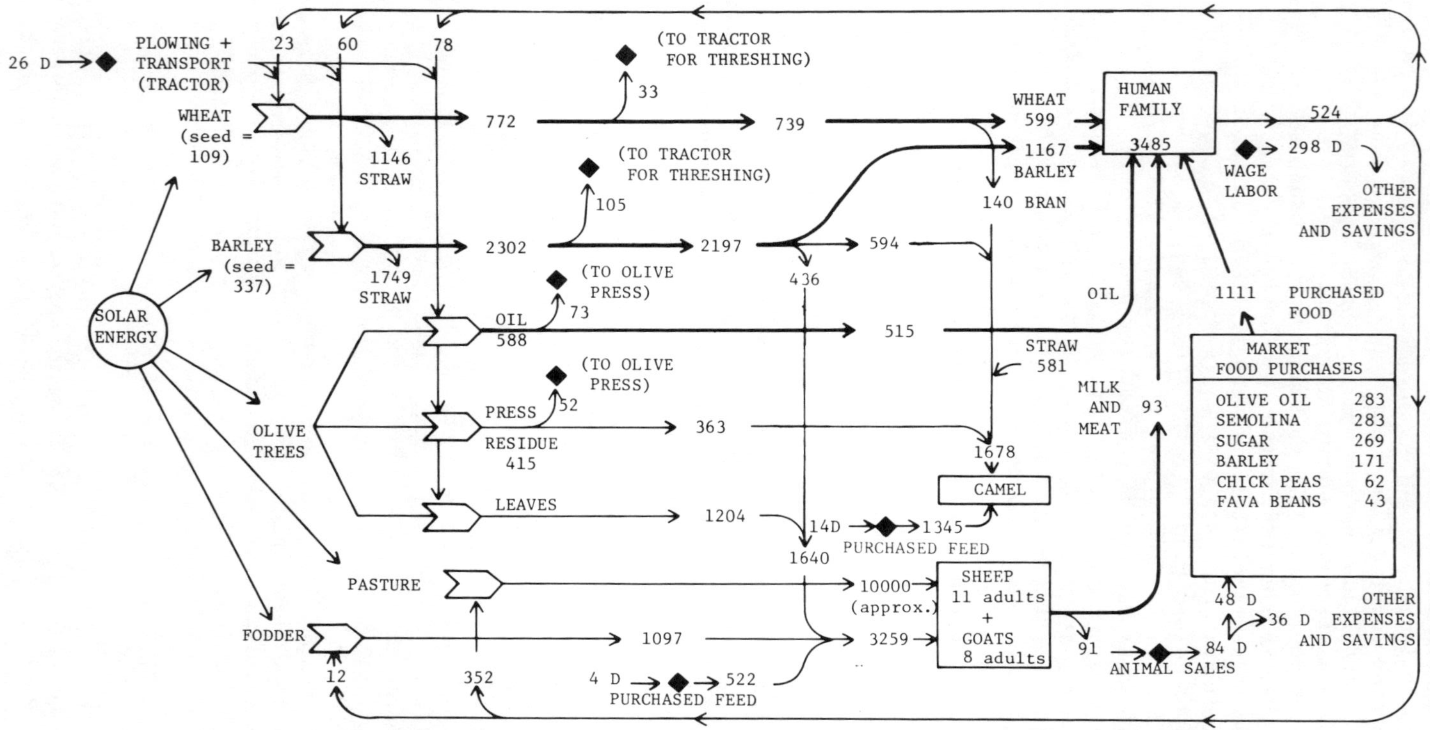

FIGURE 4. ENERGY FLOW (DURING 1973-74) THROUGH THE HUMAN POPULATION IN THE PRESAHARAN ECOSYSTEM

the short length of the subsequent growing season, the grain crop resulting from this late planting was poor. Total rainfall during the growing season amounted to 85mm. The average is 100mm for this ecosystem.

Grain production. As shown in Figure 4, the grain produced in 1973 - 74 was substantially less than the normal amount consumed in a year. Large purchases of semolina helped to offset the effects of the poor wheat harvest. The amount of barley produced was also supplemented by purchases for family consumption as well as for the herds. The preceding year (1972 - 73) had been very dry and apparently all previously existing reserves had been used to support the herds in that year.

There was 33% less barley and wheat seed planted in 1973 than in 1974. There was also a 50% reduction of cash expenditure for tractor plowing services compared to 1974. These trends reflect the diminished emphasis on grain planting as a result of the low rainfall. Families are less willing to invest in either seed or tractor services when they expect a drier than normal year. The poor rainfall in 1973 also forced families to look farther from their home base for some of their planting sites. The greater distance causes some inconvenience to the family. Recently, due to increasing availability of tractors, they have sometimes been willing to alleviate the inconvenience of plowing distant sites by engaging a tractor.

Olive production. Although rainfall was less in 1973 - 74 than in the next year, the olive crop was much better than one year later. As previously mentioned, the hot winds in spring have a confounding influence on the effects of rainfall on olive crop yields. Perhaps these winds were not severe in 1973 - 74. There was slightly less emphasis on tractor plowing services in the olive orchards during the drier year than during the following year. This was compensated by an increase in plowing with a draft animal.

Olive production in 1973 - 74 was still not at a level which could support normal family and herd needs during the year. The amount of oil purchased during the year was 35% of the amount consumed by the family. Olive meat press residue, which remains after the oil extraction process, is used as an animal food. All of the press residue fed to goats and sheep and 60% of the residue fed to camels in 1973 - 74 was purchased.

Fodder harvest. The amount of annual herbs and perennials harvested in spring for use as summer fodder was 50% below the 1974 - 75 level. This was not enough to support the supplementary diet needs of the sheep and goats during the summer of 1974. One million kilocalories (kcal) of fodder and two million kcal of other types of feed were used to supplement the diet of goats and sheep. Perhaps 20% of the total annual dietary needs of the sheep and goats were met by food not derived from grazing. However, most of this supplement was produced in some way on the home base and only a minimal amount of feed was actually purchased (less than 5% of the total diet).

Cash income and expenses. Data on animal sales were not collected for years prior to 1974 - 75 because it was felt that the family head would not be able to recall earnings which occurred in previous years. Animals are sold by lots of two or three during the entire annual cycle, and unit prices change with the seasons. It was felt that such a pattern would be difficult to recall after a new annual cycle had already begun. In order to arrive at an estimate of income from animal sales prior to 1974 - 75, the family head was asked about the number of adult female goats and sheep in his herd during each of the two previous annual cycles. Adult females were counted because they produce newborn male animals which are the main source of income from animal sales. Total cash earnings from lamb and kid sales in 1974 - 75 was divided by the number of adult females owned in 1974 - 75. This factor (4.3 Dinars per adult female) was multiplied by the average number of adult females in the herd for the years prior to 1974 - 75. In this way income from herd sales was estimated for 1972 - 73 and for 1973 - 74.

During the year each family earned an average of 298.4 Dinars in wage labor income. Animal sales produced as estimated 84 Dinars, so that the total cash income was 382 Dinars. During the same year, routine cash expenditures amounted to 97.6 Dinars. A large part of the income was saved, probably as insurance against the effects of a future dry year. At the end of the agricultural cycle in 1973 - 74, the family had little or no grain, oil, or animal feed reserves. The family would therefore have been in a poor position to absorb the effects of a dry year if it had not had a cash reserve.

Economic Production During a Year of Low Rainfall (1972-1973)

Rainfall during the 1972 - 73 agricultural year was a mere 15mm in the study area. By local standards, this was not enough to justify a large-scale grain planting effort on the home base land. Nevertheless, several families planted a small amount of grain on the home base with poor results. One third of the sample families planted no grain at all during the year. The remaining families planted grain only at distant locations where rainfall had been better. Thus, in spite of an extremely low amount of rainfall on the home base, many families were able to produce a modest crop by cultivating distant land.

Grain production. The schematic representation of energy flow during the 1972 - 73 year is illustrated in Figure 5. The grain crop produced was the smallest of all three years for which data are available. However, purchased barley amounted to only 19% of the total amount of barley consumed by the family. The quantity of wheat purchased was also minimal (2% of all wheat consumed), but this was supplemented by purchases of semolina during the year.

No additional seed was purchased for the barley and wheat planting of 1972. As noted above, family heads are less willing to risk the loss of seed and labor on a poor harvest which can be anticipated during a year of low rainfall.

Another result of the tendency to minimize risks in a bad year is the decreased use of tractors for plowing. Tractor costs for planting grain in 1972 - 73 were the lowest of any year observed (5.8 Dinars). However, more money was spent on tractor transport of the harvest to the home base in 1972 - 73 than in 1973 - 74. This increase reflects the greater distance of the planting sites from the home base during a dry year in the study area.

Olive production. Questions on olive production during 1972 - 73 were deleted from the economic survey in order to incorporate questions concerning a bumper year of olive production (1969 - 70). Production during this earlier year is not reported here. According to some informants, olive production was minimal on the home base in 1972 - 73. The estimate for olive production in Figure 5 has been set lower than the 1974 - 75 level to reflect informants' statements. It is probable that large purchases of olive oil were necessary to supplement family production.

Fodder harvest. The amount of annual and perennial plants harvested in spring 1973 was quite comparable to the amount collected in 1973 - 74. This similarity is

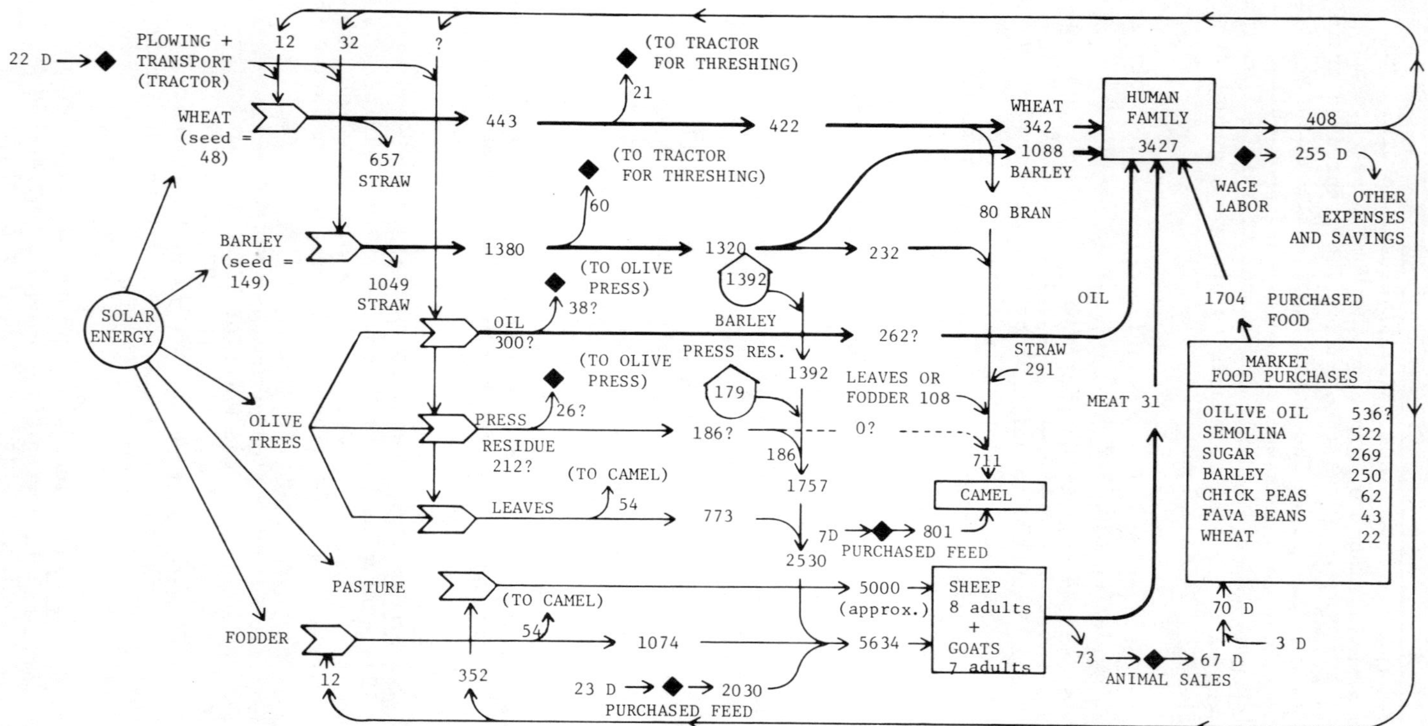

FIGURE 5. ENERGY FLOW (DURING 1972-73) THROUGH THE HUMAN POPULATION IN THE PRESAHARAN ECOSYSTEM

somewhat surprising, since rainfall was so poor on the home base. This may indicate that in 1972 - 73, the family collected plants from distant land holdings where rainfall was sufficient to support adequate vegetation growth. The one million kcal of harvested fodder was not sufficient to fulfill the food supplement needs of the herd in 1972 - 73. Two million kcal of purchased feed was also given to the goats and sheep. In addition, two and one half million kcal of feed from home-produced agricultural products were given to the herd. The total non-grazed portion of the sheep and goat diet amounted to more than 5.6 million kcal or 50% of the estimated caloric need of the herd in 1972 - 73. The purchased feed component was 20% of the total diet of sheep and goats.

Animal diet supplementation versus herding transhumance. Since the inhabitants of the study area were formerly more nomadic, one might expect that during a dry year on the home base the herds would be moved, as in the past, to an area of better rainfall. This would reduce the need for feed supplementation. However, in 1972 - 73 only one of the 24 study families moved the herd entirely away from the home base. Three others moved the herd to a different location for only six months during the year. Therefore, most of the animals in the study area were on the home base during the dry year.

As indicated above, the amount of dietary supplementation in 1972 - 73 was considerable. In the past, a dry year on the home base would always have meant a move of the herd and the family to better pastures. At present there are two choices which the family will consider during a dry year. The herd can be maintained on the home base and supplemented with non-grazed animal feed. Alternately, it can be sent with a professional shepherd (or a responsible family member if one is available) to a better pasture area.

Most of the families in the study area chose to keep the animals on the home base during 1972 - 73. This involved a considerable expense in the form of cash expenditures for purchased feed and of large quantities of home-grown feed. The amount of money spent by the sample families on purchased feed in 1972 - 73 for the sheep and goats was 22.7 Dinars. If these same families had instead chosen to send their animals with a shepherd for even six months, the cost would have been at least 40 Dinars. This cost is based on the standard herding service fee of .4 Dinar per adult animal per month. The cost of sending the animals away

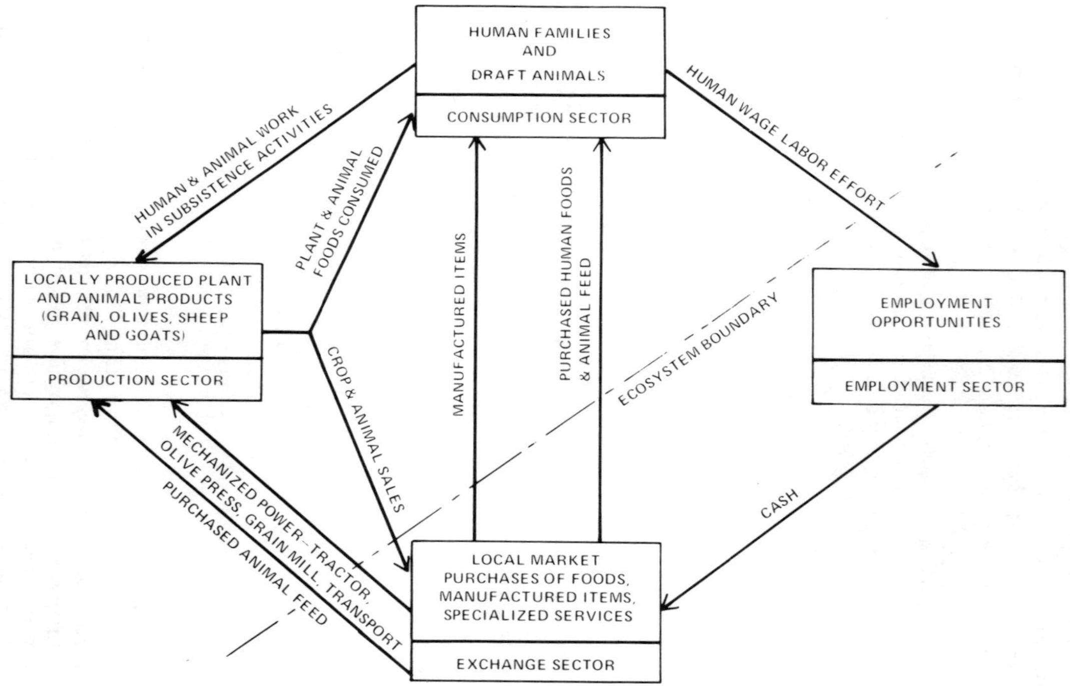

FIGURE 6. SIMPLIFIED ENERGY FLOW SYSTEM INTEGRATING WITH A CASH ECONOMY

from the home base during a poor year is therefore substantially more than the cost of leaving the animals to graze on the home base in combination with a dietary supplementation program.

Cash income and expenses. The method of estimating cash income from animal sales in 1972 - 73 is the same described previously. There were on the average, 15.6 adult females reported in the herds for 1972 - 73. The estimated cash income from sales of animals was 67 Dinars. In addition to this, 255 Dinars were earned from wage labor. A total cash income of 322 Dinars was available to the family during that year.

The routine family expenses during 1972 - 73 amounted to 79 Dinars. However, this figure cannot be compared to the expenses of other years because data are lacking for production and expenses related to the olive harvest. However, a rough estimate for these expenses is 42 Dinars, based on corresponding data for the other years studied. Adding this to the actual documented expenses, the total was 121.6 Dinars.

This estimate is larger than the amount for any other year studied. This reflects the generally poor quality of the agricultural cycle and the resulting large expenses of food purchases for family and herd. The total cash income reported above is the lowest of any of the years studied. Therefore the net surplus cash available for savings is lower in 1972 - 73 than at any other time for which data are available.

Dependence on the Ecosystem For Population Support

The information on annual energy flow can be averaged over the three years for which data are available. This procedure should provide an approximate indication of average economic conditions during the recent past. The result can then be evaluated to determine whether the population can meet its needs from food produced and cash earned exclusively from within the ecosystem. Figure 6 will help to clarify how a population which needs cash for the purchase of services, goods, and foods not obtainable in the ecosystem can, in spite of this, be thought of as being independent of sources of income from outside of the ecosystem.

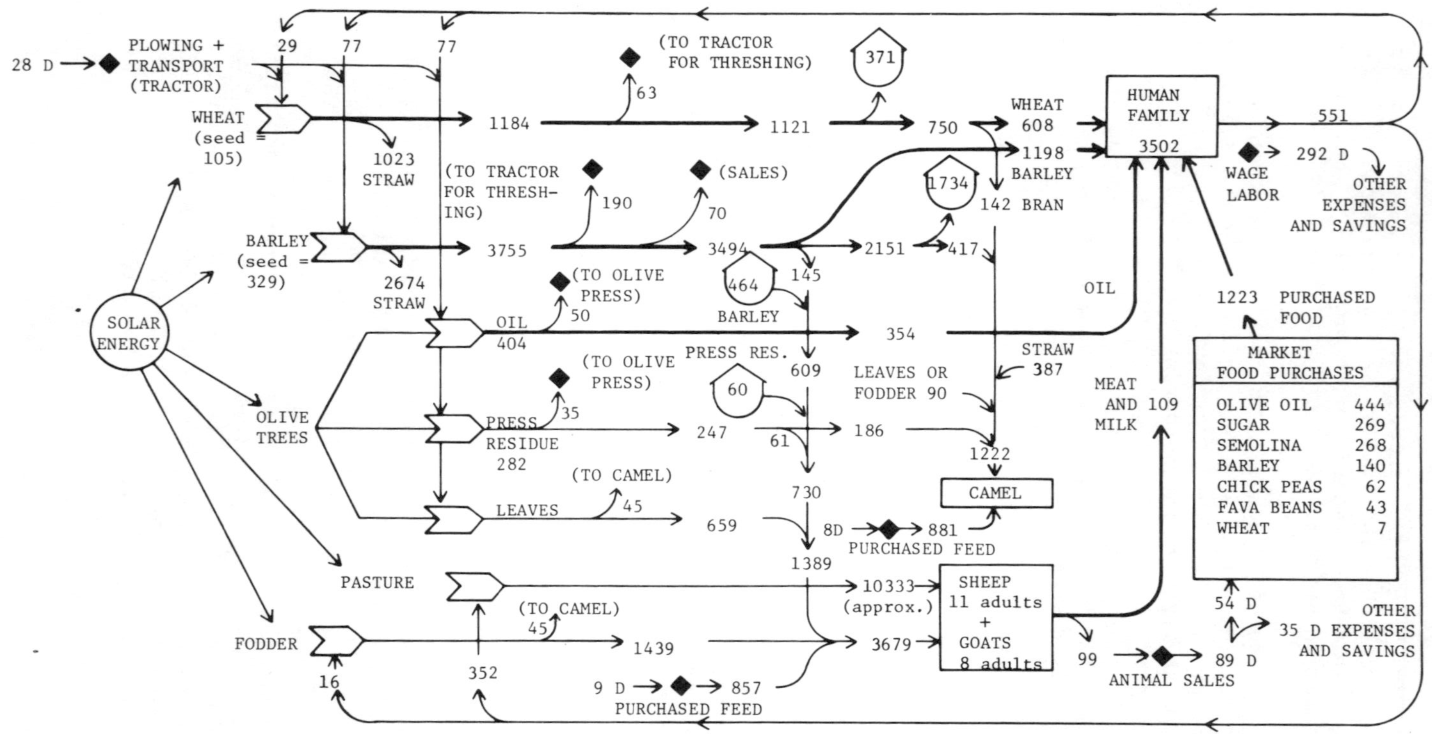

FIGURE 7. ENERGY FLOW (THREE YEAR AVERAGE) THROUGH THE HUMAN POPULATION IN THE PRESAHARAN ECOSYSTEM.

Figure 6 is a much simplified representation of the average energy flow through the human population in Southern Tunisia. If family needs are to be completely satisfied by the proceeds of economic activities performed within the system, then two conditions must be met:

1. The dietary needs which cannot be met directly from the production sector must be met by the sale or exchange of local products in order to obtain food items produced in other ecosystems.

2. There must be sufficient cash from the sale of unconsumed local products to enable the family to purchase outside services and manufactured goods. This situation could be represented by modifying Figure 6 so that there were no links to the employment sector. Wage labor opportunities would simply be ignored by rural families since they would earn enough income from the sale of locally produced goods to purchase all the specialized goods and services they need.

By averaging the data on energy flow during the three year period (Figures 3, 4, and 5) one can determine if the two conditions mentioned above can be met. If they can, then production is sufficient to meet the requirements of consumption and exchange of local products by the population. There are some conceptual problems in analyzing the average energy flow conditions based on the three years for which data are available. The order in which the wet and dry years occurred in the study period is important in terms of how one views overall energy balance for the three years. In this three year period a poor year preceded a nearly average year, which was followed by a very abundant year. During the poor year there was a heavy emphasis on purchasing food for human and animal consumption. During the near-average year there was still a large amount of purchased food in the human diet. During the abundant year an excellent grain crop was produced, enabling the family to put a large amount of grain into storage for anticipated periods of shortage in the future. When these three years are averaged together (Figure 7) a peculiar result is obtained. From Figure 7 it can be seen that an average net wheat production of 1121 thousand kcal was obtained over the three year period. This amount is slightly higher than the 1088 thousand kcal of wheat needed by the family each year. One might assume at first glance that the South Tunisians produce enough wheat on the average to meet their dietary needs. However, if the next step in the flow chart is considered, one finds that an average of 371

TABLE I. ANNUAL NET ENERGY BALANCE FOR GRAIN IN THE PRESAHARAN SUBSISTENCE SYSTEM. QUANTITIES EXPRESSED IN 1000'S OF KILOCALORIES AND BASED ON FIGURE 7.

		WHEAT
	+ 1121	home production
	- 750	consumption of home-grown wheat
=	+ 371	annual surplus put into storage
	+ 371	
	- 7	purchased wheat
	- 268	purchased semolina
	- 70	bran--loss from conversion of annual surplus into usable human food
	- 105	seed for subsequent planting
=	- 79	ANNUAL NET ENERGY BALANCE
		BARLEY
	+ 3494	home production
	- 1198	human consumption of home-grown barley
	- 145	home-grown barley to goats and sheep
	- 417	home-grown barley to camel
	- ?	consumption by donkey or mule, chickens, dog
=	+ 1734	annual surplus put into storage
	+ 1734	
	+ 70	sale of home-grown barley
	- 150	purchased barley for camel
	- 171	purchased barley for goats and sheep
	- 140	purchased barley for family
	- 464	barley taken from stored reserves of home-grown product for goats and sheep
	- 329	seed for subsequent planting
=	+ 550	ANNUAL NET ENERGY BALANCE

thousand kcal of wheat are put into storage each year. Considering that nearly all the average wheat production (1088 of 1121 units) should be used in the human diet, such storage does not make sense. In addition it can be seen that an average of 268 thousand kcal of semolina (a substitute for whole grain wheat in the diet) were purchased. This also seems inappropriate since the extra purchase should not have been necessary if the family had enough wheat to meet its average needs.

The peculiarity can be explained by recalling the three year production sequence. Abundant overproduction occurred in the last of the three years and could not possibly have been used two years earlier during the year of wheat shortage. If the abundant year had been the first year in the series, then the surplus could have been used later during the poor year. In that case, no extensive purchasing of grain products would have been observed.

How, then, should the average energy flow of agricultural products be evaluated? One method is illustrated for wheat and for barley in Table I. Since different patterns occurred in each of the years studied, the average energy flow situation represented in Figure 7 will include an average amount of grain put into storage, an average amount of grain taken out of storage for consumption, and an average amount of grain purchased to supplement family and herd needs. All three components will occur in the "average" year represented in Figure 7. In evaluating net energy balance under these conditions, grain produced locally is assigned a positive value, while grain consumed locally is assigned a negative value. Any grain or grain product imported into the ecosystem (e. g. purchased in the market) is also assigned a negative value. The method used in Table I is to calculate the average amount of home-grown grain which is stored (average local production minus average human and herd consumption of home-grown product). From this amount one subtracts the average amount of grain taken from storage (if any) and the average amount of purchased grain (or its equivalent). This manipulation is done in terms of kilocalories. The remaining amount of food energy should represent the average net accumulation or deficit in grain during the three year study period. In this way the effects of the order in which wet and dry years occur are eliminated.

From Table I it is evident that for wheat there is a slight deficit in net annual energy balance. However, this deficit is rather small. It only amounts to 6.6% of the annual need for wheat for human food and for seed. The errors involved in the estimates in Table I probably amount to more than 6%. Therefore it is inappropriate to accept this small deficit as a proven fact. The safest conclusion is that the population is very nearly in energy balance for wheat over the three year study period.

Calculating the net energy balance for barley involves some additional problems. There are several consumers in the family economy which eat barley and which have not been considered in the energy flow model. These are the family's donkey or mule, chickens, and dog. Quite clearly the combined nutritional needs of all these animals is far less than the nutritional needs of the human and animal consumers which have been considered in the model. Thus the majority of energy consumed by humans and domestic animals has been accounted for. However, the small amount of unaccounted energy is important because of the size of the apparent net energy surplus for barley reported in Table I. It is not a large surplus and could be substantially diminished if an estimate was available for the average annual barley needs of these minor consumers. Research efforts are presently being directed towards estimating barley consumption of these consumers. The safest conclusion based on present knowledge is that the population is probably self sufficient in barley production but that there may not be any net accumulation of barley in this system.

The validity of Table I as an estimate of annual energy balance is based on two assumptions. The first is that the average rainfall during the three year study period (98 mm) was representative of the long term average in the PreSaharan area (100 mm). This appears to be a valid assumption. Therefore the amount of grain produced can be considered to be a realistic average for the PreSaharan ecosystem, since grain production is a function of rainfall.

The second important assumption on which Table I is based is the constancy of the food consumption rate (kcal/year) of the various consumers in the family and its herd. Is this consumption rate (based on a three year average) representative of the long term conditions for families in this ecosystem? Family consumption during a given year must equal caloric requirements which are determined by demographic composition and body biomass of the family and the herd. The average herd size increased during the three year study period. This increase may have been a maximizing strategy to increase lamb and kid sales by decreasing older adult female sales and increasing young adult female purchases during years of progressively better rainfall. The extra investment in purchased females would have been considered safe during 1974 - 75 because of the expected high level of vegetation production for that year.

During the poor year, selling strategies could not be observed. If the three year study period had been more arid, perhaps adult female animals would have been sold, and fertility of females would have been lower. Increased sales and decreased fertility would have diminished the natural rate of increase of the herd. On the basis of available data no conclusion can be reached concerning the stability or growth of herd size. It is not known whether the three year increase in herd size will continue over the long term in Southern Tunisia.

For the human consumers, no data are available on changing family size or composition. Based on present data for the human and non-human consumers, it is impossible to assess the validity of the assumption that the family economy during the three year study period shows a rate of energy consumption which will remain the same over the long term. Perhaps a reasonable way to deal with the uncertainties involved with Table I is to conclude simply that during the study period, the family seems to have been able to produce enough grain to meet the needs of all the consumers associated with it. Returning to Figure 6, it can be said that the production sector of the energy flow system can support the consumption sector in terms of the need for grain.

This balance of local production and consumption is not true for olive oil. From Figure 7 it can be seen that an average of 404 thousand kcal are produced each year. Reported family consumption of olive oil is 798 thousand kcal per year. Therefore a large amount of oil was purchased each year from sources outside the ecosystem. This suggests an annual energy deficit for olive production in the ecosystem.

The next step in assessing the overall productivity of the ecosystem is to determine if there is a sufficient amount of unconsumed local products which can be converted to cash to enable the family to purchase needed food, manufactured goods, and services. By far the most important of these local products are sheep and goats sold from the herd. Certain categories of animals are sold in the local market, thereby converting ecosystem production into cash via the exchange sector. The average cash income from the sale of these animals is 89 Dinars each year. It must be spent via the exchange sector to provide foods and manufactured items for the consumption sector. The cash must also be used to provide mechanized power services and purchased animal feed to support the production sector. The average cash needs of the consumption and production sectors are 103 Dinars each year. This is based on accountable expenses during the three year study period. This amount is probably an underestimate of average annual cash expenditure. Immediately it is apparent that the 89 Dinars derived from unconsumed food resources in the production sector cannot pay for all the cash needs of the family (103 Dinars or more).

It is evident that this population cannot produce enough cash-earning products to support its need for money. It has already been shown that it cannot produce enough olive products to meet its dietary needs. Therefore neither condition 1 nor 2 as specified above are met by economic production in this ecosystem. As a result, the human population is dependent upon outside sources of income to meet its dietary and cash needs. Thus Figure 6, (which includes the link to the employment sector) represents the flow of energy in the study population.

Participation in wage labor by individuals from this population is explained as a result of the demonstrated need for sources of income from outside the ecosystem. An average of 292.6 Dinars was earned from wage labor each year by sample families. This labor was performed in industrialized areas outside the PreSaharan ecosystem. The estimated 103 Dinars of average annual expenditures has been previously shown to be an underestimate of family expenses. This partly explains why the total income (292.6 Dinars from wages and 89 Dinars from animal sales) is so much greater than the estimated cash needs of the family. Another possible explanation is that the families are attempting to earn a cash surplus to develop a large family treasury. Several possible motivations for this have already been discussed.

Annual Differences In Soil Disturbance From Human Economic Activity

Part of the purpose of the research in Southern Tunisia was to determine the human contribution to the processes of soil and vegetation deterioration called desertization. Just as economic production varies with the amount of annual rainfall, so does the amount of land put into cultivation. The specific method of cultivation used to make the land productive also varies with the amount of annual rainfall. Specific grain harvesting methods depends on the quality of the harvest and these also differentially influence soil disturbance.

Olive Production

The most important characteristic of the olive tree as a productive resource is that it is the only stationary resource in the South Tunisian economic system. Each year the olive orchards must be cultivated at least twice, regardless of the amount of rainfall. There is no other productive resource for which repeated cultivation is necessary. The effect of periodic plowing can be seen in Figure 8. In the foreground a mobile sand dune can be seen indicating that loose sand is accumulating down wind from the olive orchard. The only aspect of olive cultivation which appears to change each year in response to the amount of rainfall is the relative balance between plowing by tractor and by draft animal. With the tractor, the depth of penetration of the plow into the soil is deeper which causes more soil disturbance. Data on the amount of tractor plowing of olive orchards is shown in Table II. In a year of excellent rainfall there is more emphasis on tractor plowing than is the case in a more average year. However, even in an excellent year, less than 50% of the orchard land was plowed with a tractor.

Grain Cultivation

Many of the major aspects of grain cultivation and production can be seen to vary with annual rainfall (Table II). The area of land put into grain cultivation increases sharply from the poorest to the best year. This is accomplished with the aid of the tractor. The apparent strategy is to plant as much grain as possible in a good year and to cut back sharply on the area planted during a dry year. This is an effective method of maximizing grain production in a good year and minimizing the potential loss of invested resources in a bad year. The result of the observed planting strategy is that the increase in harvest size in a good year is proportionately greater than the increase in area planted or than the increase in effort, money, and seed invested.

An illustration that the above is an expectation of the family is the willingness to invest large amounts of money in tractor plowing services during a year of above average rainfall. The same trend is seen for grain cultivation as for olive cultivation. Namely, during an abundant year there is more emphasis on tractor cultivation than during an average year. During the abundant year (1974 - 75) tractors plowed 66% of the land put into grain production. The

Figure 8. Well plowed olive orchard. Mobile sand dune in foreground indicates unstable soil surface in orchard. Distance between trees is 25 meters.

TABLE II. VARIATIONS IN COMBINATIONS OF LAND USE TECHNIQUES (SAMPLE AVERAGES) AS A FUNCTION OF ANNUAL RAINFALL DURING THREE AGRICULTURAL CYCLES. DATA FROM VERBAL QUESTIONNAIRES

	(1972-73)	(1973-74)	(1974-75)
annual rainfall on home base*	15 mm	94 mm	186 mm
<u>OLIVE CULTIVATION</u>			
olive orchard area plowed by tractor	---	1.5 ha	2.1 ha
olive orchard area plowed with draft animal	---	4.1 ha	3.6 ha
<u>GRAIN CULTIVATION</u>			
total area planted in grain	3.7 ha	8.5 ha	12.5 ha
total grain area plowed by tractor	2.0 ha	4.0 ha	8.2 ha
total grain area plowed with draft animal	1.7 ha	4.5 ha	4.3 ha
money spent on tractor for grain planting (Dinars)	5.8	8.0	16.4
wheat yields (kg/ha)	122	94	197
barley yields (kg/ha)	158	114	253

*Measured on Tunisian PreSaharan project study site (average of 5 rain gauges).

Figure 9. Wheat stubble remaining after harvesting with sickle.

proportion of tractor plowing during the other years is close to 50%.

An important variable mentioned earlier, which is also determined by annual rainfall on the home base, is the distance of the grain planting sites from the home base. In a wet year a large portion of the grain planted by the family will be concentrated on the home base. Some grain is also planted on distant sites, however, in order to maximize the anticipated harvest size. On the other hand, in years of lower rainfall on the home base, the family must plant all or nearly all its grain at more distant sites. In those years the only tractor cultivation on the home base is the minimal amount in olive orchards. The concentration of tractor plowing on the home base during a wet year is a practice which is likely to increase desertization.

Techniques for harvesting grain also vary. If the wheat harvest is good and the plants dense, they are cut with a sickle. This leaves a stubble which is grazed by the sheep and goats, leaving the roots in the ground to help stablize the soil surface and decrease wind erosion (Figure 9). If the wheat harvest is sparse due to poor rainfall, the entire plant is pulled with the roots. For barley, regardless of the quality of the harvest, the entire plant is usually pulled. Since barley covers the major part of the land cultivated in grain (70%), its characteristic harvesting technique is the dominant factor determining the contribution of grain cultivation to soil disturbance. Most of the time, barley patches are left completely denuded after the harvest (Figure 10).

Impact of Land Use Types on Desertization

In order to assess the importance of the major land use types (grazing, olive culture, grain culture) as factors contributing to desertization, two variables must be considered. One is the amount of land area subjected to a given land use type, and the other is the amount of soil disturbance per unit area. In terms of the amount of area used, grazing has the greatest impact on soil disturbance. Based on sample averages, 54 hectares are grazed on the home base each year by the family herd. Only seven hectares are cultivated in olives, and an average of eight hectares are cultivated in grain (averaging wet and dry years together). In terms of overall surface area alone, one would conclude that grazing is the most important contributor to vegetation destruction, soil movement, and desertization.

Figure 10. Barley field after harvesting by hand. No stubble remains after the harvest.

The impact of sheep and goat grazing can be seen in Figures 11 and 12. In Figure 11, the study site of the Tunisian PreSaharan project is shown in an oblique view from the air. The enclosed borders of the study site are clearly visible due to the contrast between the thick vegetation cover within the site compared to that found on the locally grazed land surrounding the site. The vegetation improvement has occurred within three years on the study site even though the vegetation is exposed to a controlled grazing regime. Figure 12 is a vertical view of a human habitation (large arrow) near the PreSaharan project study site. A part of the study site appears as a darker area in the corner of the photo. The animal corral (circled) is surrounded by a light area (arrows) which is completely lacking in vegetation cover. This results from the constant trampling of the vegetation by the animals milling around the corral site. Mobile sand dunes accumulate in a semicircle (dashed line) down wind from the corral.

In terms of the amount of soil disturbance <u>per unit area</u> grazing is not the most important contributor to vegetation destruction. Obviously plowing creates more disturbance per unit area than does grazing. As mentioned above, olive cultivation is always performed several times per year on a given piece of land. This does not happen with grain cultivation.

In terms of the frequency and degree of soil disturbance per unit area, one can rank various land use types in terms of their contribution to desertization. Olive cultivation would rank highest as the most destructive because of frequent plowing. Home base grain cultivation would be intermediate, and distant-land grain cultivation would be the least destructive since plowing is not frequent on a given site. In this ranking, grazing would be the least destructive of the land use types per unit of area disturbed.

Combining the two land use characteristics (per unit soil disturbance, and number of hectares disturbed) one can get an idea of the overall importance of each land use type as a contributor to desertization. Grazing covers 78% of the land used by the family and plowed land amounts to 22% of such land. The least destructive land use type (grazing) covers the largest portion of exploited land, and the most destructive land use type (plowing) covers the smaller portion of exploited land. In this sense, it appears that grazing and plowing make approximately equal contributions to the processes of desertization.

Figure 11. Oblique aerial view of the study site of the Tunisian Pre-Saharan Project. The fenced borders of the study site show clearly due to the increased vegetation growth within the enclosure. A large part of the site is subjected to a controlled grazing program.

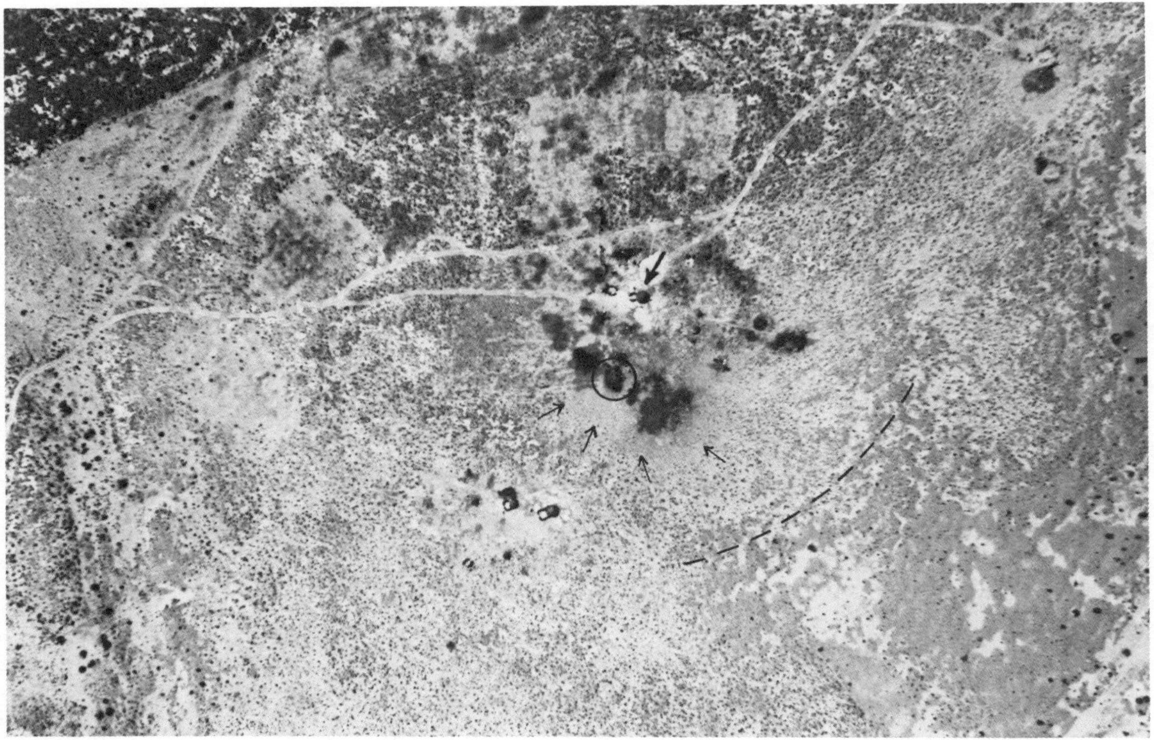

Figure 12. A "halo" of denuded terrain surrounds the corral and human habitations in the center of this photograph. See text for explanation.

Land Use Strategies In Southern Tunisia

It has been shown that energy flow in the human population in Southern Tunisia is at a level which is slightly below the minimal level necessary to support the human and domestic animal population. This is probably a recent development resulting from decreased land availability due to a steadily growing human population. In the past, more land was available to each family, herd sizes were larger, and animal sales provided a larger cash income. It is likely that in the past the population was independent of sources of income originating outside of the ecosystem. Under those conditions, energy flow may have been a limiting factor on the human population. One would expect to see population responses which maximized energy production or minimized energy expenditure. These patterned behavioral and/or biological responses may have remained in the present system. If they have, one would expect that food production would yield a high energy return on the investment of human labor.

Efficiency of Agricultural Production

The efficiency of subsistence production can be evaluated in terms of the amount of human effort required to produce the crops and animals which serve the food requirements of the population. An effective or efficient method of subsistence production is one which minimizes energy expenditure in work effort and maximizes energy production in crops. This efficiency is always viewed within the context of the constraints of the ecosystem. Given a set of possible subsistence alternatives, it is expected that individuals will choose those methods and crops which result in the highest long-term energetic efficiency (food energy produced/work energy expenditure).

Compared to other low-productivity ecosystems the energetic efficiencies of South Tunisian subsistence crops seem to be fairly high. They are as follows:

1. Barley grain 46.3 : 1
2. Wheat grain 38.6 : 1
3. Olive oil 7.7 : 1

Using barley as an example, the meaning of this ratio is that 46.3 kcal of grain are produced for every kilocalorie spent as human work necessary to produce the crop. These estimates are rather approximate for several reasons.

First, animal food such as chaff, bran, straw, olive leaves, or olive-press residue has not been considered in the production factor. Second, the amount of draft animal effort is excluded from the expenditure factor. Third, the money earned in animal sales has a caloric work cost and, since some money is spent in crop production, the portion of this money resulting from animal sales has a caloric cost. This cost should be considered in the expenditure factor of the efficiency ratio. Data are currently being analyzed so that these items can be included in the efficiency ratios of the South Tunisian crops.

The estimates show that the grain producing system is well adjusted to the constraints of the ecosystem. Olive oil production is not an efficient process compared to grain production. However, this ratio does not take into account the production of animal food from olive by-products. The efficiency ratio would be closer to 15 : 1 if all materials were considered in the production factor. This complex of efficiently produced crops is what one would expect if energy flow was formerly a limiting factor for this population.

Conversion of Surplus Animal Production into Alternate Foods

One of the main subsistence strategies of the PreSaharan population is to convert animal production into cash to purchase foods and goods and services which are not locally available. The conversion of live-animal energy to cash to purchased food energy involves an energy return which is a useful way of evaluating this pathway of energy transformation. Table III provides the various rates for the conversion of live animal energy into purchased food energy via the market system. The indicated amount of meat energy will bring one Dinar at the market (live animal sales). One Dinar, in turn, will buy the specified amount of any of the indicated food alternatives.

The conversion of live animal energy into human food energy via the market system brings a considerable calorie return to the family. A return ranging from 15 to 50 kcal of human food energy is received for every kilocalorie of live meat sold. In terms of subsistence strategies it can be said that the animal herd acts as a bank of energy reserves of which a small part is converted into human food every year. Considering the kinds and average amounts of foods purchased (based on Figure 7) there is a 25 : 1 overall return of human food energy purchased for every kilocalorie of animal meat energy sold.

TABLE III. CONVERSION OF GOAT AND SHEEP PRODUCTION INTO OTHER FOOD ITEMS VIA THE MARKET SYSTEM (VALUES BASED ON MARKET CONDITIONS IN 1974 - 1975).

CONVERSION OF ANIMALS INTO ALTERNATIVE HUMAN FOODS

		57,500 kcal barley*
		41,000 kcal flour
		38,400 kcal wheat, semolina
1094 kcal meat =	1.0 Tunisian Dinar =	29,400 kcal pasta or sugar
		21,000 kcal chick peas
		20,000 kcal fava beans
		16,600 kcal olive oil

AVERAGE: 1094 kcal = 1.0 Dinar = 27,529 kcal purchased food

AVERAGE ENERGY RETURN FROM ANIMAL SALES: 25:1

CONVERSION OF ANIMALS INTO ALTERNATIVE ANIMAL FOODS

		192,000 kcal olive leaves
		190,000 kcal press residue
1094 kcal meat =	1.0 Tunisian Dinar =	100,000 kcal hay
		93,000 kcal bran
		64,000 kcal barley*

AVERAGE: 1094 kcal = 1.0 Dinar = 96,000 kcal purchased feed

AVERAGE ENERGY RETURN FROM ANIMAL SALES: 87:1

*The price of barley for animal food differs from the price of barley for human food.

The conversion of animals into purchased animal feed is also included in Table III for comparative purposes. It is interesting that there is a higher energy return if animal food is purchased rather than human food. A return ranging from 58 to 175 kcal of animal food energy can be purchased for every kilocalorie of live meat sold. In monetary terms it is less costly to purchase animal food than human food. Based on actual feed purchase patterns, there is a 87 : 1 overall return of animal feed energy bought for every kilocalorie of animal meat energy sold.

Economic Aspects of Food Production Versus Food Purchase

Since the energy conversion is so high from live animals sold to food purchased, one might wonder if there is any energetic or economic advantage at all for home produced food compared to purchased food. It has already been shown that the energetic returns on home produced food are high. There is also an economic advantage to the home production of food compared to the purchase of food. Home production yields more energy per Dinar spent than would be the case for outright purchase of food. This is because a family which produces its own crops only spends its available cash on incidental agricultural costs such as payment for tractor services or purchase of seed for planting. In Table IV the total energy value of the average crop harvest has been divided by the total cash expenditure for incidental agricultural costs for the production of that crop. In effect, the crop is "bought" with a certain amount of cash expenditure and a certain amount of family's labor expenditure. The resulting values in energy produced per unit of money spent are presented in Table IV. Labor energy has not been deducted from the energy value of the crop. A comparison with the values in Table III shows that much more energy is obtained for each Dinar spent in home production than is the case for market expenditures on food. If the energy returns on cash inputs alone are considered, the family can obtain five times more grain per Dinar by growing its own, and it can obtain twice as much olive oil per Dinar by home production. The obvious result is that a family has the same amount of energy available at a lower monetary cost. The energetic and economic advantages of home production are apparently a sufficient inducement to the continuation of subsistence agriculture.

TABLE IV. ENERGY RETURNS FROM AGRICULTURE PER UNIT OF MONEY SPENT.

Barley	294,459 kcal/Dinar
Wheat	212,280 kcal/Dinar
Olive Oil	30,456 kcal/Dinar

Short Term Strategies: Maximizing Output and Minimizing Risks

Land holding and inheritance. Although the sampling zone depicted in Figure 2 is rather restricted, the families living within the zone own and use land which is scattered over a 75 km strip running roughly parallel to the coast. This strip also runs parallel to the rainfall isohyets. There is also some indication that the vegetation composition of these sites tends to be somewhat uniform. Thus a rather large but specific ecological microzone is considered appropriate and necessary for exploitation by local families. The sample families own sites scattered throughout the strip in small parcels of less than five hectares each. Five or more of these small plots may be owned. This scattered pattern may be explained in terms of the striking degree of spatial variability in annual rainfall in the PreSaharan zone. For instance, the small sampling zone (150 square kilometers) does not uniformly receive rainfall within its boundaries. During the fall of 1972, no measurable rainfall was recorded by the rain gauges on the PreSaharan project study site. Yet two kilometers east of the rain gauges, grain planting was attempted in that year. This must have meant that at least 10 - 15mm of rain fell on the eastern site or the people would not have been convinced that planting was worth the risk. This indicates a sharp difference in rainfall over a short distance.

In addition to the acute spatial variability in amount of rainfall, there is no predictable pattern as to the specific location of rain in a given year. Each part of the PreSaharan zone is equally likely to receive rain. As a result of the unpredictability in amount and location of rainfall, the scattered land holding pattern has a distinct advantage. Such a pattern maximizes the probability that adequate rain will fall on at least one plot of the family's available land in each year. This is the best assurance the family has that it will have access to arable land each year.

The nature of inheritance rules in this population assures that the scatter of land holdings will be maintained between generations. When a family head divides his land, <u>each</u> son gets an equal share of <u>each</u> plot of land. If a family head had five pieces of land scattered throughout the zone, each of his sons would receive five smaller pieces scattered in the same way. The nature of the scatter is maintained. The problem of continually decreasing sizes of each generation's land holdings is partly balanced by temporarily borrowing land or by purchasing land adjacent to the small plot which is owned. Purchase is the most likely solution for lands adjacent to the home base, and borrowing is the expected solution in the case of distant land holdings.

<u>Optimizing strategies of agricultural production</u>. If the land holding system fails to dampen the effects of the extreme annual variation in site-specific rainfall, then the amount of cultivation is adjusted to correspond to the expected rainfall. Inhabitants of the PreSaharan zone determine the amount of the first autumn rainfall by measuring the depth to which moisture has soaked into the top layer of sand. There is a lower limit of autumn rainfall below which people will not plant.

In addition to the above there is another means of interpreting autumn rainfall. If the rainfall is of substantial amount and occurs early, this is a clue that the rest of the agricultural cycle will also be moist. This is a predictable characteristic of PreSaharan rainfall. The amount of autumn rain is directly proportional to the rain which will occur during the balance of the growing season. This fact is well recognized by the local inhabitants. They make judgements about the adequacy of autumn rainfall as well as predictions of potential crop yields for the ensuing harvest. Based on these expectations the family adjusts the amount of money, seed, and labor it will invest in a crop each year. During a year which is expected to be very moist, every effort is made to plant grain over as large an area as possible. The investment is high but the risks are considered low. During a year which is expected to be dry, little or no planting occurs. The investment is very small because the risks of crop failure are perceived to be very great. This kind of planning explains the very great range in seed, cash, and labor investment illustrated for the three year period in Table II and in the energy flow models for each year (Figures 3 - 5).

Strategy Changes Imposed by Increased Sedentarization

During the past 20 years, new water sources have been added along the main highway (Figure 2) which serves the PreSaharan zone. This has been accomplished by the construction of a pipeline by the Tunisian government. The pipeline originates in an oasis north of Medenine whose rate of water flow is capable of providing water all along the pipeline route. This new water source has attracted settlement in this part of the region. Families which formerly owned a small parcel of land near this road and water source have decided to make this parcel their permanent home base. Concrete houses have been built and adjacent land has been purchased. Dependable water sources are very important, especially during the summer when every goat and sheep in the region must receive water every second day. This consumption periodically put such a drain on the previously used hand-dug wells that seepage of ground water into the well was occasionally surpassed by human and animal consumption of its water. During the summer, people may have to wait several hours for seepage to refill the well to a minimal level which enables water to be taken by bucket. The dependability of the new pipeline water source is one of the main attractions to permanent settlement near it.

This water source and the mobility provided by the road have encouraged permanent settlement in a five kilometer strip on either side of the road. There is presently a commitment to sedentary residence which was not evident in previous generations. As recently as 15 years ago a family or part of a family would move to the location most suited to grain production and herd grazing during each new agricultural cycle. This had the effect of spreading the destructive aspects of land use over a wide area, with longer fallow periods for vegetation recovery between each use at the same site. Presently, individuals occasionally find themselves with productive land (due to good rainfall) at their disposal which is too far from the home base to be exploited effectively. The result is that the home base may be cultivated even though rainfall has been better on a distant parcel of family land. The home base is used more frequently with a resulting decrease in length of the vegetation recovery period. This type of intensification of land use speeds up the processes of desertization.

When the inhabitants were more nomadic in their settlement patterns they were more likely to move to distant areas for grain planting. The people are now more reluctant to move draft animals, plows, seed grain, human and animal food, and items of shelter to distant grain planting sites than previously. To a moderate extent the need for the mobility involved in distant-site grain planting has been eliminated by another new strategy. When people wish to plant grain on a distant site they engage a tractor to carry the seed grain to the locale and to plow the land. This can be easily accomplished in one day. Formerly the family would have been obliged to endure a two week stay in the distant area. Alternately the family might have chosen to remain at the new location for the whole year if it had access to sufficient grazing land for the herd. At present, because of the availability of tractor services the family head may depart for a distant area and return home the next day after watching the tractor plow the area of land which would otherwise demand two weeks of his time. This eliminates the need to transport the family and the materials necessary to plow a distant site.

Prospects for Developing Land Use Alternatives

Current land use practices have a clear and undeniable detrimental effect on vegetation and soil stability. These effects are exacerbated by the recent trend to settle in permanent dwellings on a home base. The home base land is used much more than it would be if traditional standards of acceptable levels of rainfall were applied. Recovery periods have shortened so that vegetation regrowth has slowed considerably. Each year more land is planted in olive orchards, with the result that more land is permanently cultivated. The orchard is the most destructive form of land use in terms of vegetation removal and soil disturbance. It is unlikely that these land-use trends can continue very far into the future and still yield a sustained level of agricultural and pastoral production.

Incentives to Maintain Current Practices

Although the destructive nature of current land use practices is regrettable, there are substantial incentives within the economic system to encourage the continuation of these practices. These have been dealt with throughout this discussion and are summarized here:

1. It is cheaper to produce as much food as possible using current methods of subsistence agriculture. One Dinar will produce five times more grain if it

is invested in home production than it would if it were spent at the market in outright purchase of grain.

2. The cash acquired from the sale of young male herd animals is presently greater than the cash spent to maintain the herd. This is true despite the considerable expense of purchased animal feed which must be used during a poor year. Thus there is a net cash income from animal husbandry.

Two additional incentives within the system have not yet been dealt with. These are outlined in some detail.

Market conditions. The retail price of olive oil has nearly doubled in the past few years. This trend is a further inducement to exploit the local system. As the price of purchased olive oil increases, the difference in returns per Dinar become even more favorable for home production (Table III compared to Table IV). Thus there is an increasingly strong incentive to plant more olives in order to produce more oil. Most of the new planting takes place on the home base.

Population pressure, herd size, and tractor use. One of the most critical incentives within the system is quite complex. It is a result of the following chain of events.

After the first major autumn rain there is a period of approximately 15 days during which grain can be planted. If it is planted after this period there may not be enough remaining soil moisture to support the sprouting and initial establishment of the grain seedlings. During this time period a man and camel can plow 5.6 hectares of land in grain (18.75 hours of planting/ha, and seven hours of plowing/day). This is based on field measurements of working rate.

In order for families to achieve the average annual grain production, 8.23 hectares were plowed each year. In Figure 13 the actual amount of plowing is shown for each of the three years. In two of the three years, the total amount of cultivated land exceeded the area which a man and camel could plow (dashed line). Most of the families in the sample own only one draft animal. Therefore the 5.6 hectares limit on hand plowing applies to most of them.

Because of this limitation on area plowed by draft animal, the subsequent harvest, based on traditional cultivation methods, would be lower than the annual need for grain. The deficit can be made up in two ways. The grain can be purchased or it can be grown by the family with the aid of the tractor. In Southern Tunisia the tractor can plow 12 times faster than a draft animal. The combination of draft animal <u>and</u> tractor plowing enables the family to plow all the necessary land within the two week sprouting period while spending only a minimum of cash on tractor services.

Formerly, herds in the PreSaharan zone were larger and tractors were unavailable for grain planting. More land was available to support the larger herds due to the smaller size of the human population in the area. The strategy in the past was to plow by draft animal as much land as possible and to make up the deficit in production by purchasing grain. This was accomplished with the proceeds from animal sales. This income was greater in the past because of the larger herd size than at present. It was therefore possible for the family to put forth the extra cash expenditure of buying grain.

At present,available land area per family has decreased and herd sizes have been adjusted downward. This has meant diminishing income from animal sales. At the same time, home production of grain with the aid of the tractor is less expensive per kilogram of grain harvest than is the purchase of grain. If one of the rules of the economic game in Southern Tunisia is to minimize cash expenditure, then the choice in recent years is to produce more grain with the aid of the tractor. This option would be used before grain would be purchased. The shift to the tractor seems to be an unavoidable consequence of local economic strategies and of the limits imposed by the working rate of the camel and man.

The results of the chain of events outlined above can be seen in Figure 13. The height of each column reflects the average amount of land planted in grain during the study period. The proportion of surface area plowed by tractor and by draft animal is indicated for each year. The number of days needed to hand-plow the area is indicated inside the part of the column which refers to hand plowing. Next to each column is a time period (in days) which expresses the amount of time which would have been needed to plant <u>all</u> the land by hand.

The important trend in Figure 13 can be seen by comparing the two latter years. Both times the family spent 12 days in hand plowing efforts. But during the abundant year

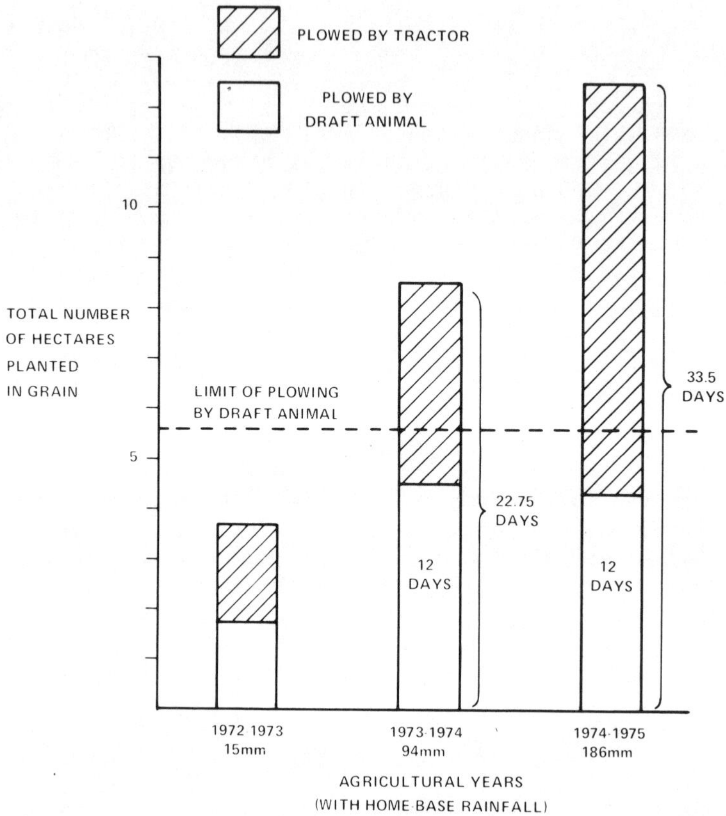

FIGURE 13. MECHANIZED AND TRADITIONAL PLOWING PATTERNS DURING THREE AGRICULTURAL CYCLES

the family used the tractor to increase greatly the area of cultivated land. In this way the family can increase the amount of grain planted during a year when the expectation is for low probability of crop failure. Based on the above, one would expect that families would have hand-plowed all the grain fields in 1972 - 73. This is because the area falls well within the 5.6ha limit which can be plowed by hand. The reason that some tractor plowing occurred is due to the recent reluctance of the family head to plow distant locations by draft animal. Figure 13 clearly illustrates the real importance of the tractor in the family's strategy to maximize grain production during years of low risk.

Evaluating Alternate Land Use Strategies

It is clear that strong incentives exist within the system to maintain the present strategies of land use in Southern Tunisia. The energy returns are high on investment of labor and cash in the local productive system. It will be difficult to find alternate land use patterns which are simultaneously less destructive and also maintain an equally high return on investment. However, it is also fairly clear that if the present intensity of land use continues on the home base, desertization will render the entire area useless for years to come.

The analysis of this ecosystem is somewhat preliminary. Ecological research by the Tunisian PreSaharan Project is still in progress. It is therefore inappropriate at present to attempt to propose land use changes which will retard or halt the ongoing process of desertization. However, based on present knowledge it is possible to suggest several characteristics of a potentially successful land use alternative. Alternative strategies or resources should be no less efficient than present ones. The energy return on investments of cash, labor, and materials should be about as high as is the case for the present subsistence complex. The level of risk involved in the investment should be no higher than the risks associated with the current resource complex.

The standard used to judge land use alternatives is the current land use system. The present system works well from the point of view of the native Tunisian, although ecologically it is rather destructive. This paper has been concerned with a description of why and how the system works from an energetic and an economic point of view. The values of energy efficiency ratios and energy returns on cash investment can be used as standards with which alternate strategies should compare favorably if they are to be acceptable to the local inhabitants over the long term. Admittedly there are other important standards of acceptability of a proposed economic change. There are social and ideological aspects which have an important bearing on the acceptability and feasibility of alternate subsistence strategies. These attributes are still under study in the South Tunisian population and are beyond the scope of this paper.

One might argue that the energy ratios and returns in Southern Tunisia are excessively high because the system does not operate on a sustained yield basis. That is, a system operating only on an immediate return basis (ignoring the consequences for desertization) may be more efficient in the short run than a sustained yield system which has no negative impact on the environment. This may be true. In that case the standard of judgment (the current efficiencies) is unreasonably high. Nevertheless, the most successful programs of economic development (from both the inhabitant's and the ecologist's point of view) will be those which can approach the energy returns of the present system while maintaining a sustained yield based on soil and vegetation stability. This standard of judgment is basically applicable to economic planning in all subsistence economies. The energy returns of a system must be determined before alternate strategies can be evaluated. A study of energy flow is uniquely capable of determining energy returns by combining energetic and economic points of view.

REFERENCES

Leung, W. W.
1968 Food Composition Table for use in Africa. Bethesda and Rome: National Center for Chronic Disease Control, and Food and Agriculture Organization of the United Nations

Little, M. A. and G. E. B. Morren, Jr.
1976 Ecology, Energetics, and Human Variability. Dubuque: Brown

Maelzor, D. A.
1965 A Discussion of Components of Environments in Ecology. Journal of Theoretical Ecology 8:160

Rapp, A.
1974 A Review of Desertization in Africa: Water, Vegetation, and Man. Stockholm: Secretariat for International Ecology, Sweden

Vidal's paper complements the previous one since they both deal with indigenous technologies. The Bedouins' accomplishment, without the labor commitment and organizing power of a state government, was on a more modest scale. But it is no less impressive than that described for Peru. Indeed, it seems a perfect example of "appropriate technology", and one which might well be reintroduced in the same and other areas.

4

Utilization of Surface Water by Northern Arabian Bedouins

F. S. Vidal

Climatologists and geographers are not always in agreement as to where the boundary lies between "arid" and "non-arid" areas in terms of measured annual precipitation, or between "arid" and "semi-arid" areas. There is no disagreement, however, when it comes to classifying the Arabian peninsula. With the exception of a small section in the southwest, which is under the influence of the monsoon winds, the Arabian peninsula is arid indeed. This vast land, the size of the United States east of the Mississippi (1), receives an average annual precipitation of only 2.75 inches (less than 75 mm.), mostly in torrential form during the rainy season (if one can call it that) between November and April. This average figure, of course, means that while there may be years of greater rainfall there are other years when the precipitation is far short of 2.75 inches. In parts of the Empty Quarter, in the Arabian southeast, there may be no rainfall recorded at all for several consecutive years. In 1969-1970 eastern Arabia recorded in some spots only 0.6 of an inch of rainfall (less than 15 mm.), in 1971 only one inch (25.4 mm.). By contrast, in 1972 northern Arabia recorded in spots almost 6.0 inches (152 mm.) of rainfall, and parts of the Eastern Province recorded over 12 inches (305 mm.) of rainfall in 1975-1976.

In Arabia there are no rivers in the accustomed sense of the word; no means to provide substantial water reservoirs of the types which are conventional elsewhere. Water problems are not peculiar to the Arabian peninsula; they are common to most of the Middle East, and particularly acute in the Sahara, and Iran, as well as in the Arabian deserts. The scarcity of water has posed a great threat to the local population in terms of sheer survival; it has also had a great psychological effect among peninsular Arabs (among whom my first-hand data were collected). Peninsular Arabs are "water-obsessed". In the Arabian environment, water--partic-

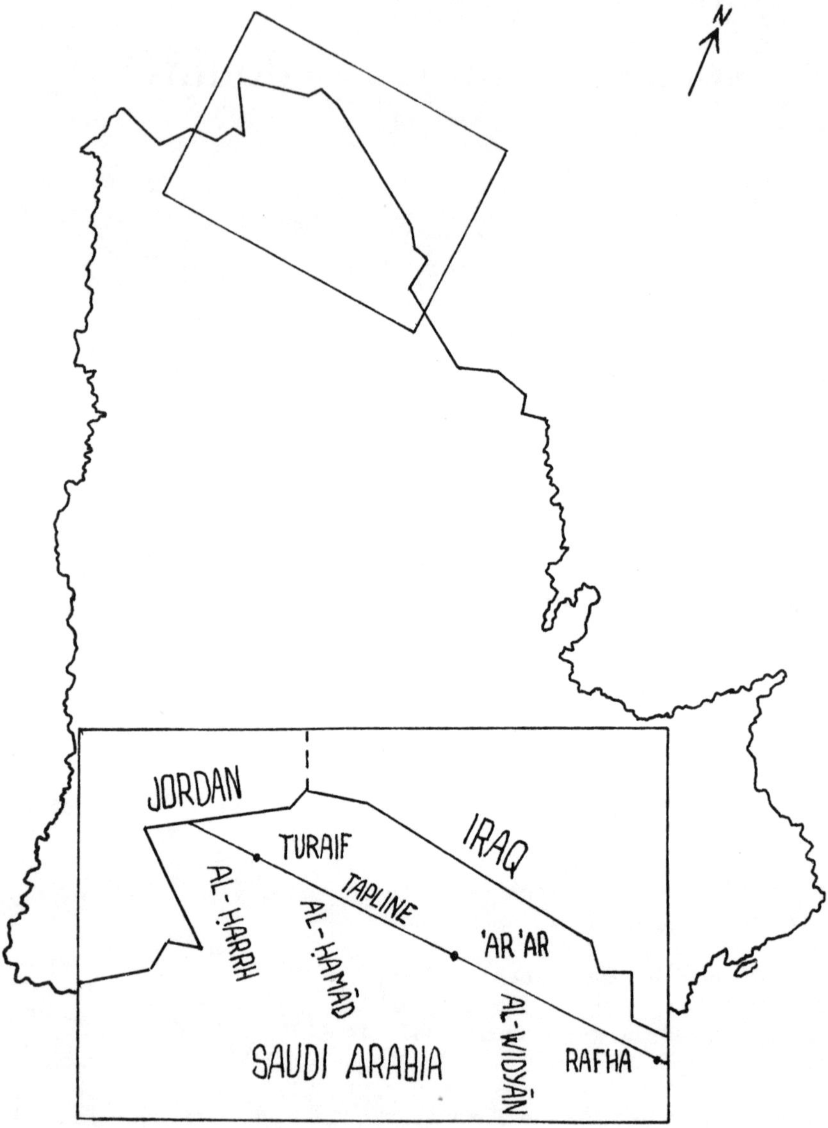

Figure 1. Sketch Map of Extreme Northwest

ularly flowing water--is such a manifest gift of God that it must not be tampered with, its flow must not be controlled. Where water occurs in abundance, as in the striking environment of such a large artesian oasis as al-Hasa in eastern Arabia (2), it has traditionally been wasted, with eventually disastrous results for the local agriculture (3). I have often remarked in casual conversation, somewhat facetiously perhaps, but with more than a little factual justification, that it was very easy to teach a desert Bedouin how to turn a water faucet on, but almost impossible to teach him how to turn it off.

In the artesian oases the availability of water is no problem; in fact, the problem is rather an excess of available water. In the open desert, however, the availability of water is critical, except in the winter, when fodder grasses are the primary concern. If the rainfall has been normal camels can derive their moisture needs from the fresh grasses and from the shallow depressions, called mishāsh, where water has collected. In the summer, when the grass has been eaten or burnt by the sun, the camel's survival, helped by the fat stored in his hump, depends on the availability of sub-surface water (4). Perhaps accessibility is a better term, since very extensive sub-surface aquifers are found in large areas of Arabia, but unfortunately at too great depths (5).

Geomorphologically the Arabian peninsula may be divided in two provinces: the older Arabian Shield (the eastern part of the Arabian-Nubian Shield) of igneous and metamorphic rocks, which (except for an eastward bulge near the center of the peninsula) forms a narrow band along the west coast; and the much larger and younger sedimentary province, which slopes gently away from the Arabian Shield towards Iraq, the Persian (or Arabian) Gulf and the Arabian Sea, and consists primarily of sandstones, limestones, and shale. In the violently upthrust Arabian Shield some peaks reach a height of over 12,000 feet; in the sedimentary province the few cuestas, mesas, ridges, and other erosion remnants are mostly of negligible relief. Along the igneous/sedimentary contact volcanic activity (some as recent as the early 1800's) has produced extensive lava fields.

The area with which we are concerned here is the extreme northwest of Saudi Arabia (see sketch map, figure 1) consisting of three geomorphologically distinct regions. These are, from west to east: al-Ḥarrah, mostly volcanic fields of basalt and andesite overlying Eocene limestones; al-Ḥamād, a moderately dissected high plateau of rolling chert plains; and al-Widyān, a highly dissected area of steeply sloping, deeply eroded wādī (6) drainages. Both al-Ḥamād and al-Widyān show mostly Cretaceous deposits.

It is interesting to note that in the gently downward sloping eastern portion of the sedimentary province, impervious layers which trap above them the rainwater seeping through the sands and other permeable layers, are frequently found relatively close to the surface, often between 20 and 70 feet (some 7 to 25 meters), an easy depth to reach with the help of a limited tool-kit; hence the many hand-dug Bedouin wells in this area. In the western reaches of the sedimentary province, on the other hand, the water table is found at great depths. In the region of our present concern water is found in two Rafha locations at 480 and 530 feet (160 to 180 meters), in two 'Ar'ar locations at 985 and 1,060 feet (330 to 350 meters), and in three Turaif locations at 865, 930, and 1,600 feet (290 to over 500 meters) (7). These are depths which cannot be reached with the ordinary hand tools in the Bedouin inventory, and to my knowledge there are no hand-dug wells in this part of the peninsula.

It is also of great interest to point out that these areas of Arabia, particularly al-Ḥarrah and al-Ḥamād, by a most unfortunate coincidence offer in years of good rainfall some of the best pasture in the country, while the critical sub-surface water remains inaccessible. Considering the above, it is not surprising that with the usually scant rainfall, and the depth of the aquifers, the people of the Middle East have for centuries--if not millennia--devoted considerable time, energy, and ingenuity to trapping surface water, wherever it could be found. The qanāts of Iran (called foqqāra in northwest Africa, and aflāj in the Arabian peninsula) are well known but need not concern us at this point since this was primarily a technique to transport underground water, and we are here interested in the collection of surface runoff. Professor Evenari and his colleagues (8) at the Hebrew University in Jerusalem have done extensive work on the utilization of runoff water in the Negev desert, getting their inspiration in part from old Nabatean practices. In Pakistan we record the technique of water-spreading low dams drawn in a zig-zag pattern (9). In Oman we find remnants of what are called gabarbands (10), low stone barrages across wādī beds, which would trap both the silt and water of torrential floods, but did not usually last very long, possibly because of an excessive gradient.

In northwest Arabia the surface drainage water was collected in maḥāfīr (sg. maḥfūr, meaning "excavation"). These maḥāfīr are remarkable structures. They consist of an open earth-work ring of variable height (generally from two feet to ten or twelve), and variable diameter (usually from fifty or sixty feet to some 300 or more), enclosing a pit of variable depth (see Figure 2). They are easily identifiable from the air, looking like a horseshoe, or a volcanic crater with a broken edge, or--to use a more pedestrian analogy--like a doughnut minus one bite. The opening of a maḥfūr always faces the direction from which the drainage water flows. The maḥāfīr were built by excavating roughly circular areas to varying depths, and piling up the excavated earth around the circular pit as a retaining wall. In some localities this retaining wall was considerably strengthened by the use of stone blocks, some of which had evenly dressed surfaces for better fitting.

There are two standard locations for a maḥfūr. The preferred one seems to be any khabrā (pl. khabārī), which is a silt flat formed at the terminal point of an alluvial drainage (11). Further drainage is prevented either by surrounding higher land, by sand dunes, or where a large body of standing water is formed during the rainy season due to negligible slope. Cobbles, pebbles, and sand would have been deposited upstream; the silt then is left behind on the terminal plain as the water evaporates. On occasion, though rarely, the term khabrā is also used to refer to silt flats in other than terminal areas of a drainage system. For example, at one point of its course a wādī may spread out over a large area of very gentle gradient, thus depositing silts before continuing along its course. At any of these points, when the water stream reaches a wide area of gentle slope it spreads out, loses velocity, is subject to high evaporation rates, and deposits a large proportion of the silt it carries in suspension.

The other preferred location for a maḥfūr is the confluence of two wādīs, where again there is a decrease in the velocity of the water and an increase in the deposition of silt. In this case the maḥāfīr are generally located on the downstream side of the confluence point and on the side of the tributary. In the khabārī the maḥāfīr are commonly built along the edges of the flat. I have seen maḥāfīr in the center of wādīs, but never in the middle of a khabrā. These locations make a good deal of sense, as otherwise if the maḥāfīr were constructed more towards the center, while they might become filled with water a little sooner, the surrounding mudflats would prevent the use of maḥfūr water for a long time, and create a very uncomfortable access. Generally, then, the locations of maḥāfīr are about as represented in figure 3.

The maḥāfīr fulfilled a double function. In the first

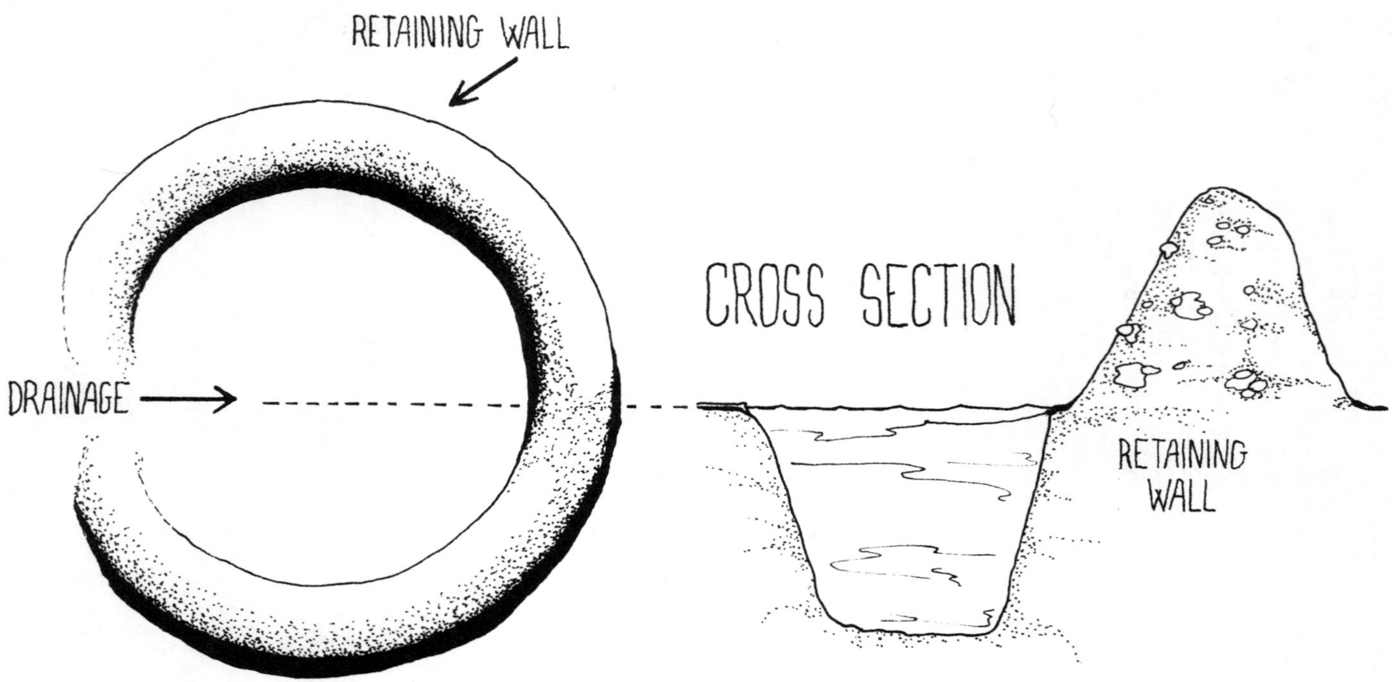

Figure 2. Cross-Section of a *Maḥāfīr*

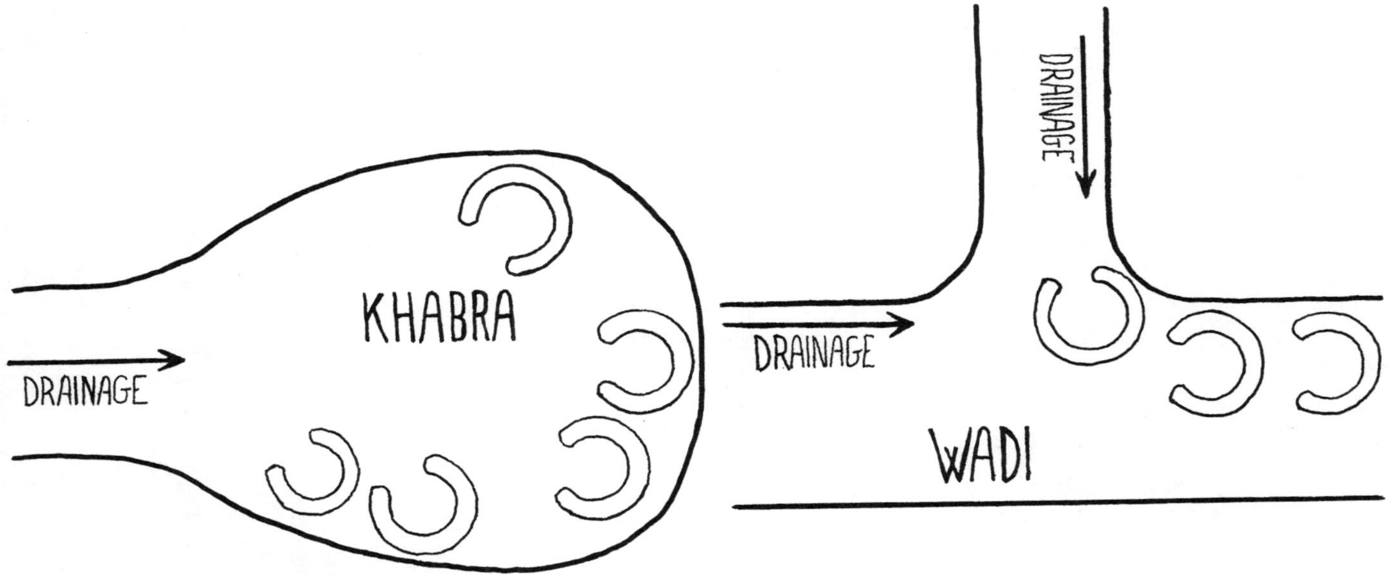

Figure 3. Location of _Maḥāfīr_

place they supplied drinking water for people as well as livestock. Secondly, they attracted the once plentiful wild game of the area, so that the vicinity of a maḥfūr was a favorite with hunters. In fact, even though flint implements can be found in many locations in the general area of northwest Arabia, any amateur "arrowhead" collector has a much better chance of rich pickings by concentrating on the vicinity of a group of maḥāfīr, which are easy to spot, rather than having to look for other localities, the discovery of which usually requires not only a longer time, but more experience and specialized knowledge. I should not like to venture a guess as to the age of these maḥāfīr structures. To call them (or at least some of them) "prehistoric", or "from the Stone Age" might be justifiable. But in the absence of proper archaeological excavation we do not know as yet what these terms may mean in Saudi Arabia. The "Maḥāfīr People", whoever they were, could have been living on the technological level of the stone age at the same time the Imperial Romans maintained a garrison in Dawqirah, near Turaif, right in the midst of maḥāfīr country, just about 2,000 years ago.

The location and construction of the maḥāfīr were perfect for their purposes; the decrease in the velocity of the water made it possible to get by with a relatively weak retaining wall (which also provided a dry approach to the water's edge); secondly, the fine silt deposited on the bottom of the maḥfūr contributed to the impermeability of the surface, so that the maḥfūr became better able to hold water; their location near the edge of the silt flat made the water accessible almost as soon as the floods had started; finally, the water depth in a maḥfūr also helped preserve water, due to lower evaporation rates compared to the surrounding silt flat, where water would stand at a depth of only a few inches and would evaporate in a matter of just a few days. On the basis of aerial photo strips I have estimated that there are 730 maḥāfīr in northwestern Arabia: 230 in al-Ḥarrah, and 500 in al-Ḥamād (12). The total of 730 is a conservative figure, since the few scattered maḥāfīr in al-Widyān are not included, and structures which were of doubtful identification were also left out. Still, considering these figures, the harvesting of surface water by the nomads of northwestern Arabia appears to have been a major undertaking.

It can be roughly estimated that the bulk of the maḥāfīr hold anywhere from around 100,000 gallons of water to about 1,000,000 gallons. Inasmuch as a camel's prodigious innards are capable of stowing away some 50 gallons at each watering (13), a maḥfūr can take care of between 2,000 and 20,000 animals. These figures are obviously somewhat high, since allowance must be made for some evaporation, seepage, and waste, plus the use of some water by other than camels.

There is one important drawback in the use of the maḥāfīr: over the years they slowly become silted in to the level of the surrounding silt flat, and hold much less or hardly any water. They are, however, easy to clean out and maintain in satisfactory condition though minor maintenance work must be carried out regularly and frequently, just as is done with the hand-dug wells in other parts of Arabia. To allow a maḥfūr to become completely silted in would mean to lose it altogether. Although some maḥāfīr continue to be used today, most of them have not been cleaned for a long time, hold comparatively little or no water, and are no longer the reliable water reservoirs they once were.

The maḥāfīr, as the hand-dug wells in other areas of the peninsula, have played an important part in local politics and local warfare. A relatively small hostile force could easily hold a maḥfūr, or a group of maḥāfīr, long enough to deny the use of the trapped water by another people over a critical number of days. A hostile force could repeat this for several years, and allow the maḥfūr to become completely silted in. Or else, the retaining wall around the structure could be wantonly destroyed, as has happened in some places. Such hostile acts could be (and probably most often were) carried out by an opposing tribe, but they could also sometimes be due to administrative action, since in this region the availability of water can become an instrument of national policy. A strong central administration could also influence tribal displacements by assigning (or by denying the use of) certain maḥāfīr to specific tribal groupings.

I have already pointed out that while maḥāfīr abound both in al-Ḥarrah and al-Ḥamād, particularly in the latter, they are practically absent in al-Widyān. I do not know the precise reasons for this, since this is an area of good grazing and no hand-dug wells. The most likely explanation may be the configuration of the terrain. Over the greater part of al-Widyān the drainage gradient is much steeper, and the consequent higher velocity of the water might have knocked down any retaining walls people attempted to build, thus forcing the experiment to be abandoned.

A possible solution in this area might have been the construction of a series of stone masonry cisterns (birkahs) which are like "de luxe" maḥāfīr. This was done around the year 800 A.D., in a region lying just to the east, when the series of birkahs on the Darb al-Zubaidah (Zubaidah's road) was built. Queen Zubaidah, the wife of the Caliph Harūn al-Rashīd (a contemporary of Charlemagne) was a very devout lady who financed the construction of a series of such water cisterns with attached caravanserais between Baghdad and Medina for the comfort of travelers on the Muslim pilgrimage. Such structures are, however, very costly to build and would most probably not be practical today.

In most recent years a different type of reservoir has contributed to the harvesting of surface water in this part of Arabia, albeit in totally unplanned and unforeseen ways. I am referring to what are called "borrow" pits, dug for operational reasons in the vicinity of the several pumping stations of the Trans-Arabian Pipe Line Company (Tapline) (14). Some of these pits, generally of a smaller size, are found all along the one thousand miles of the pipe line, but the majority, and especially the larger ones, are in the vicinity of the pump stations. Most of these pits were dug in connection with the construction of the pipe line between 1947 and 1950. I know of only one which is attributed by the local population to their own efforts at an earlier time. This pit is called by the local people a *maḥfūr*, and is thus labeled on several maps, although it is not a *maḥfūr* of the classic type.

After Tapline became operational and the local towns in the vicinity of the pump stations started their mushrooming growth, other pits were dug in the general area of the towns and used for quarrying mud or gravel. Most Tapline pits have not experienced any changes since construction times, but those in the areas adjacent to pump station towns have often been considerably enlarged by local contractors in mud or gravel quarrying operations.

Many borrow pits are quite large, up to several hundred feet long and wide, and eight or more feet deep. They are generally located on low ground, and up to several million gallons of water may collect in them after the winter and spring rains. One such pit in 'Ar'ar, for instance, can take care of the watering needs of 14,000 camels on a four-day watering pattern (i.e., some 3,500 camels a day) for a period of about six weeks (15). Again, this figure is somewhat optimistic, since I have made no allowance for seepage, evaporation, waste, or water drawn for people or for other livestock, and I have assumed the bottom of the pit to be of uniform depth, which is not the case. Conservatively, this pit would take care of 10,000 camels under the same conditions, as corrected (16).

Totally unplanned for the purpose, these pits were a

godsend for Tapline during the 1950's and 60's. In some of the years of least rainfall the 'Ar'ar pit held water through August, the Turaif pit through September. Had the water not accumulated in the borrow pits, the situation in 'Ar'ar would have been extremely precarious, with thousands of Bedouins and their tens of thousands of livestock relying on an insufficient amount of water drawn from Tapline's deep wells (17).

There is at least one such pit in the vicinity of each of Tapline's pump stations, and the locations are well known by the population of the region. After they have become filled, they are all extensively used both for human consumption and as watering places for livestock. At the time of my field-work the only exception was Turaif, where the local government had exercised control to prevent the watering of animals in such pits, and had diverted them to the deep-well troughs. The water in the Turaif borrow pit was reserved for the drinking and cooking needs of the people. The reason for this is the same as that underlying a statement made by one Bedouin, "Water from the sky is better than water from the ground." Pit water, being new rain water, is much sweeter than the water produced from the deep wells in these regions of Arabia. Deep-well water in this area is of a varying degree of salinity: 938 ppm solids in Shu'bah, 670 ppm in Juraybah, 766 in Wādī 'Ar'ar, 581 in Qaryatain (18). These are, to be sure, certainly potable levels; but rainwater is very much sweeter. In any event, whether used to water animals or exclusively for the needs of the human population, the water in these pits relieved some of the pressures exerted on Tapline's water supplies (all of which were derived from deep-wells) during the most trying part of the year (19).

We thus have two facts on record: one is the old Arab development of surface water with the aid of the maḥāfīr, and its successful practice over many years, and the other is the experience Tapline has had with its borrow pits. I should like to suggest the possibility that this Arab experience could probably be successfully applied in other parts of the world where there is a need to harvest surface water, particularly in unpredictable drought years.

We might examine some of the pros and cons of such a project. The first point to be considered, namely that of cost, would seem to be favorable. It was estimated in the early 1960's that one deep water well in the 'Ar'ar area would cost about $50,000 to complete. Repair, maintenance, and operation would also be expensive. Furthermore, the drilling of such a well involves bringing in special equipment and personnel.

It was difficult to estimate the cost of mechanically digging a borrow pit, but technicians calculated that, depending on the distance of the pit from the pipe line road, the degree of hardness of the silt pan, and the size of the proposed excavation (assuming average proficiency of the operators and average condition of the mechanical equipment, and without major breakdowns or interruptions), one could be dug for between $2,000 and $6,000 (20). For the lowest figure we assumed a pit 300 feet long by 200 feet wide, and 10 feet at its deepest, located within a few hundred yards of the Badanah pump line station, and dug in the fairly soft soil of the *Wādī* Badanah bed. A pit of this size, which is smaller than some of the borrow pits in existence, even allowing for seepage, evaporation, waste, a certain amount of water drawn for other uses, access ramps for livestock, and a very gentle gradient from one side so as to allow the heavy equipment to operate faster, would still be able to hold enough water (conservatively estimated) for some 40,000 camels, or some 2,500 camels a day on a four-day watering pattern for about two weeks. This is about the same number of camels that can be watered daily from a 90 gallon/minute deep water well operating 24 hours a day at a much higher cost. The operating costs of the pit would be nil. Maintenance would consist as a rule only of de-silting every few years; it could be done with the help of heavy equipment in about one day, and would cost somewhere in the vicinity of two or three hundred dollars (21).

It would seem that if a successfully time-tested method of development of surface water is also by far the cheapest, there should be no further objections to its application, but a number of other considerations enter into the picture. During my fieldwork some people mentioned the possibility of the borrow pits becoming breeding grounds for the malaria mosquito. This is unlikely; in the first place, the pits would only hold water for part of the year, and secondly their locations would be well known and easily accessible so that control--if deemed necessary--would be easy. The bacteriological level of the water would have to be considered, but water-borne bacteria could not last a whole year, since in the Arabian desert the reservoirs would be completely dry during several months.

Another objection was that the strong odors emanating from such large herds would be wafted over campsites. This could be overcome by locating the pits downwind and at a fair distance from human settlement. The objection that large herds of livestock would be accompanied by even larger swarms of flies is perhaps the most valid one, particularly if one considers the tenacity--and the apparent immortality--of the Middle Eastern fly. In my personal experience, however, the appearance or disappearance of large numbers of flies has never been correlated with the arrival or the departure of camel herds.

A much more weighty objection to open pits, transcending the mere nuisance value of flies and odors, is the circumstance that such pits might constitute hazards to both human and animal life, and that the agencies responsible for the excavations might be held legally liable for losses through drowning. This is the reason behind the practice of draining all borrow pits after construction in the United States (a practice which is legally enforced in many States). However, other people may feel differently about the possible hazards, and agencies attempting to harvest surface water in this fashion would be well advised to ascertain a priori the local attitudes, as well as the existing liability laws (22).

In any case, and in summary, it seems to me that in these Arabian examples we may find yet another--both ancient and modern--method of harvesting surface water in other suitable arid parts of the world, to parallel the Nabatean-inspired efforts to Professor Evenari and his colleagues in the Negev.

Notes

1. This is equal to the combined areas of California, Nevada, Utah, Colorado, New Mexico, Arizona, and Texas; or about eight hundred times the area of Rhode Island.

2. This oasis has a combined total water discharge in excess of 200,000 gallons per minute from over seventy artesian springs, allowing the cultivation of more than four million palm trees. Vidal, 1954.

3. Vidal, 1955:187.

4. In the winter a camel may go for six weeks without watering; in the intense heat of an Arabian August (perhaps 120° F. in the non-existing "shade") camels must be watered every third or fourth day; sheep and goats must be watered daily.

5. Sladic, 1966:2-3. More than 200 trillion barrels. The best summary of water development in Saudi Arabia is in P. Beaumont, 1977.

6. The Arabic word wādī strictly means "river", but since over most of the Middle East rivers are rare, the term most often refers to "river bed"; the term "arroyo", used by American geomorphologists, is the closest translation.

7. Data from Aramco's Exploration Department.

8. National Academy of Sciences, 1974:23ff.

9. National Academy of Sciences, 1974: fig. 17.

10. Hastings, et al., 1975:11.

11. Most of the drainage in the Arabian peninsula is interior drainage, the sort that forms the flats which American geomorphologists refer to as "playas".

12. In reading these strips to arrive at the estimate I was assisted by D. A. Holm, an Aramco geologist.

13. In the summer a camel is taken to the water at about sunset, when he drinks some 35 gallons; he then beds down for the night, and is led to the water early the next morning to take on an additional fifteen gallons before being taken out to graze.

14. In the parlance of the construction trades in the U.S.A. these pits are generally known as "borrow" pits. In the frequently atrocious transliteration of various southwestern dialects, construction-site reports also spell the name "burrow", "burro", and even "boar". I do not know how the term originated.

15. This location could be improved to take care of 30,000 camels on a four-day watering pattern over the same period of time.

16. In Nancie L. Gonzalez' remarks on my Denver paper, she stated that she had not, before that time, thought of measuring water in terms of "camel-watering" units. Neither had I, and I am most grateful for the idea.

17. A somewhat complicated legal point must be mentioned here, about which I am not competent to pass judgment. Article XV of the Tapline Convention required that Tapline "develop" or "increase" the local water supply, and there is a fine point of distinction

when these terms are put into Arabic. The article also made reference to the "local" population, not making it clear whether what was meant by "local" referred to people long settled in the area--hardly worth counting--or people who had become "local" because of Tapline's attraction, or even temporary travelers who happened to be "local" on a transitory basis. The numbers of people involved could grow indefinitely. If Tapline had to take care of an indefinite number of people, because of an "open-ended" interpretation of its obligation it would have resulted in a very onerous financial load, given the cost of the few existing deep-wells drilled strictly for Company operations. The possibility of harvesting surface water became, once again, as in maḥāfīr times, very attractive.

18. Data from Aramco's Exploration Department.

19. The situation is somewhat different today. Tapline is a marginal operation, and Government rigs have tapped low aquifers of sweet water in a number of locations.

20. These are figures I obtained from my Tapline colleagues; considering the year (about 1960), they should be corrected.

21. These figures have not allowed for inflation. See fn. 20.

22. While I was researching the topic of this paper I interviewed His Reverence Shaykh Ṣāliḥ al-'Anqari, the Chief Judge for the Northern Frontiers Province of Saudi Arabia on this matter of legal liability with reference to Tapline, which at that time had been the only agency responsible for excavating the largest borrow pits in the area. His answers may not reflect a consensus of his fellow judges, but were nevertheless most interesting. His Reverence made the following points: 1) The Pipe Line Company had the operational need, and by its agreements with the Government, the right to dig pits; 2) These pits were "public facilities", in that their locations were well known by the people, and that they were used by the people; 3) It was public knowledge that deep water always represented a danger to animals, as well as to people who could not swim; 4) In the absence of underground seepage, the presence of water in open pits was due only to rainfall. This

water was controlled by God alone and was not Tapline's responsibility, hence His Reverence would not hold the Company liable for drowning losses, even though it had been the excavator of the pits.

When I suggested that, regardless of liability, Tapline might still like to drain or to fill in the pits because the thought that a human life might be lost was a distasteful thing for the Company to contemplate, His Reverence countered by saying that tampering with God's own water was at least as distasteful to him.

References Cited

Beaumont, P.
1977 "Water and Development in Saudi Arabia". The Geographical Journal, 143: 42-60.

Hastings, A., J. H. Humphries, and R. H. Meadow.
1975 "Oman in the Third Millennium B.C.E.", in The Journal of Oman Studies, 1: 9-57.

National Academy of Sciences.
1974 "Runoff Agriculture", More Water for Arid Lands, Washington pp. 23-37.

Sladic, M.J.
1966 The Wasi'ah Aquafer. Symposium of the Institute of Petroleum. Bahrain.

Vidal, F.S.
1954 "Data Culture in the Oasis of al-Hasa", Middle East Journal, 8:416-428.

Vidal, F.S.
1955 The Oasis of al-Hasa. New York, ARAMCO.

It takes the special skills of one trained in both anthropology and in botany to ferret out and interpret evidence concerning the use of plant materials in times past. Contemporary argicultural ethnocentrism becomes of special concern when we realize that earlier, less scientifically sophisticated peoples utilized a much greater variety of highly nutritious plants than we do. Given the impending shortages in the world food supply, it behooves us to consider ways to make the deserts bloom. Other papers have discussed this in relation to water. Felger and Nabhan suggest that we may also do it by nurturing and cultivating plants adapted to arid, semi-arid, or even saline conditions.

5

Agroecosystem Diversity: A Model from the Sonoran Desert

Richard S. Felger and Gary P. Nabhan

Two-thirds of the land surface of the earth is arid or semi-arid. As modern agriculture has expanded into dry lands to meet the world's food needs, it has not only imposed a small set of crop species and technologies on diverse traditional systems, but it has had to simulate tropical and temperate environments *en masse* to support these crops. The environmental and economic costs of supporting ecologically maladapted modern industrial agricultural crops are great: high petrochemical inputs, extravagant water importation projects, fossil ground water pumping, increasing soil salinity, and depletion of soil fertility. Millenia-old locally adapted germ plasm and techniques for sustained microenvironment exploitation have been lost during this century as a result of exotic agriculture.

To counter the inherent drawbacks in monoculture, particularly in dry lands, we suggest diversifying agro-ecosystems by increasing 1) the number of species (species richness) and 2) growth-form (life-form) diversity of agronomic crops. Based on data from the Sonoran Desert region, we feel that a variety of agronomically unexploited environmental niches could be utilized for propagation of economic plants with minimal supplements of water and energy. To maximize productivity and minimize costs, crops would be developed to fit the environment rather than the environment being modified to fit the crop. While the present discussion centers on food plants, similar concepts can also be applied to non-food crops and fauna.

Increasing the number of species utilized as food resources would, in itself, have profound nutritional, cultural, and ecological ramifications. Two general trends have been associated with worldwide shifts from hunting-gathering economies to subsistence cultivation, and finally to industrial agriculture: 1) increase in human population, and 2) overall decrease in the number of plant species utilized as food resources. These trends have progressed to the situation of present day industrial agriculture in which only a handful of plant species stand between the majority of humanity and starvation. These crop species include wheat, rice, maize, barley, common bean, soybean, and common potato. The remaining major world crops consist of not more than several dozen species. We have reached a state of extreme specialization with regard to food - our primary energy source. Are we becoming overspecialized?

Compared with the various major food plants of the pre- and non-industrial world, the major industrial world crops fall within a relatively narrow span of ecological tolerances and growth-forms. Most are herbaceous plants with relatively high water requirements, comparatively low salt tolerances, and which thrive best in actual or simulated temperate or tropical environments.

What are the worldwide potentials for increasing the number of crop species? The total number of edible species is not known. We will use ethnobotanical information from the Sonoran Desert to demonstrate a methodology for estimating the number of species which have been used, as well as the number which have served as major human food resources in the world as a whole.

The Sonoran Desert covers 310,000 km^2 and supports a flora of about 2500 species of seed plants (Shreve and Wiggins 1964). Our present data indicate that about 18% of this flora or about 450 species (about 375 native species and about 75 naturalized species) were utilized for food by the various native peoples in this region. More than 10% of the 375 edible species have been utilized as major food resources. Similar relationships are found for the plants utilized by the Seri Indians of Sonora, Mexico, who occupied one portion of the Sonoran Desert (Figures 1 and 2). The total flora of the major portion of Seri territory is estimated to comprise 500 species (for species-area relationships in this region see Felger and Lowe 1976). At least 80 species within this region, more than 15% of the total flora, are known to have been utilized for food by the Seri, and more than 10% of these (8-10 species) served as major food resources (Felger and Moser 1976). Extrapolated to the world as a whole, this allows a crude estimate of 30,000 species of edible plants, with approximately 3,000 potential major food resource species. These relationships are summarized in Table 1.

The estimate of 30,000 species of edible seed plants worldwide is probably conservative, as is our record of approximately 375 species of edible plants native to the Sonoran Desert. A more accurate estimate of the total number of food plant species worldwide as well as the number of major food resource species will depend on correlating ethnobotanical and floristic data from other, diverse regions of the world. The present crude estimate of 3,000 potential major food resource species may be in error within its order of magnitude, and the chance that any one of these species may become an important new agronomic crop is probably low. Yet if only a small fraction of these are successfully introduced into modern agriculture as major food resource species, there could be a marked improvement in the world nutritional, agricultural and ecological situation.

The major present-day agronomic crops were first domesticated by so-called Neolithic societies. However, some previously unexploited species currently being domesticated by agricultural scientists may increase the inventory, e.g. the buffalo gourd *(Cucurbita foetidissima)* from southwest North America (Bemis et al 1975). Using jojoba or goatnut *(Simmondsia chinensis)* as an example, we can roughly estimate that \$2 to \$5 million may be sufficient to introduce a new crop into modern agriculture (Wade Sherbrooke, personal communication). Research and development costs for 1,000 species of potential major new food crops might thus be on the order of \$2 to \$5 billion. This is certainly not too expensive considering the potential lasting value to humanity and the biosphere.

We suggest that increasing the growth-form diversity of crops may allow the efficient use of previously underexploited ecological niches (see Felger 1975, Felger and Nabhan 1976). For instance, agriculture has yet to penetrate much of the area adjacent to the 30,000 km of shoreline along hot coastal deserts of the world. Tidal flats and other saline soils could be managed or farmed for the production of certain halophytes such as the perennial grain, *Distichlis palmeri*. Eelgrass *(Zostera marina)* can be grown in shallow sea water in protected bays and inlets (Felger and McRoy 1975). Many other perennials could be grown with minimum tillage methods leading to substantial conservation of agricultural energy inputs (Phillips and Young 1973).

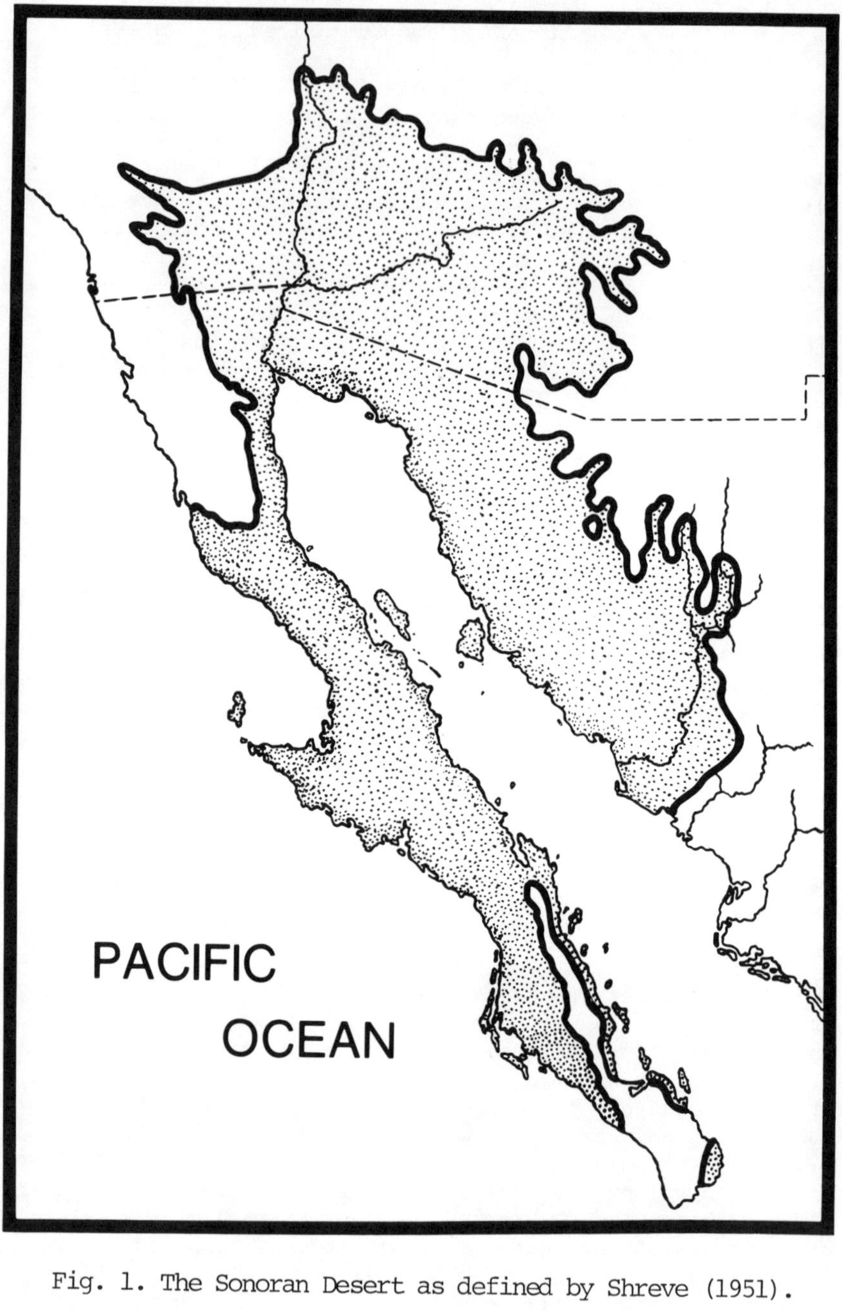

Fig. 1. The Sonoran Desert as defined by Shreve (1951).

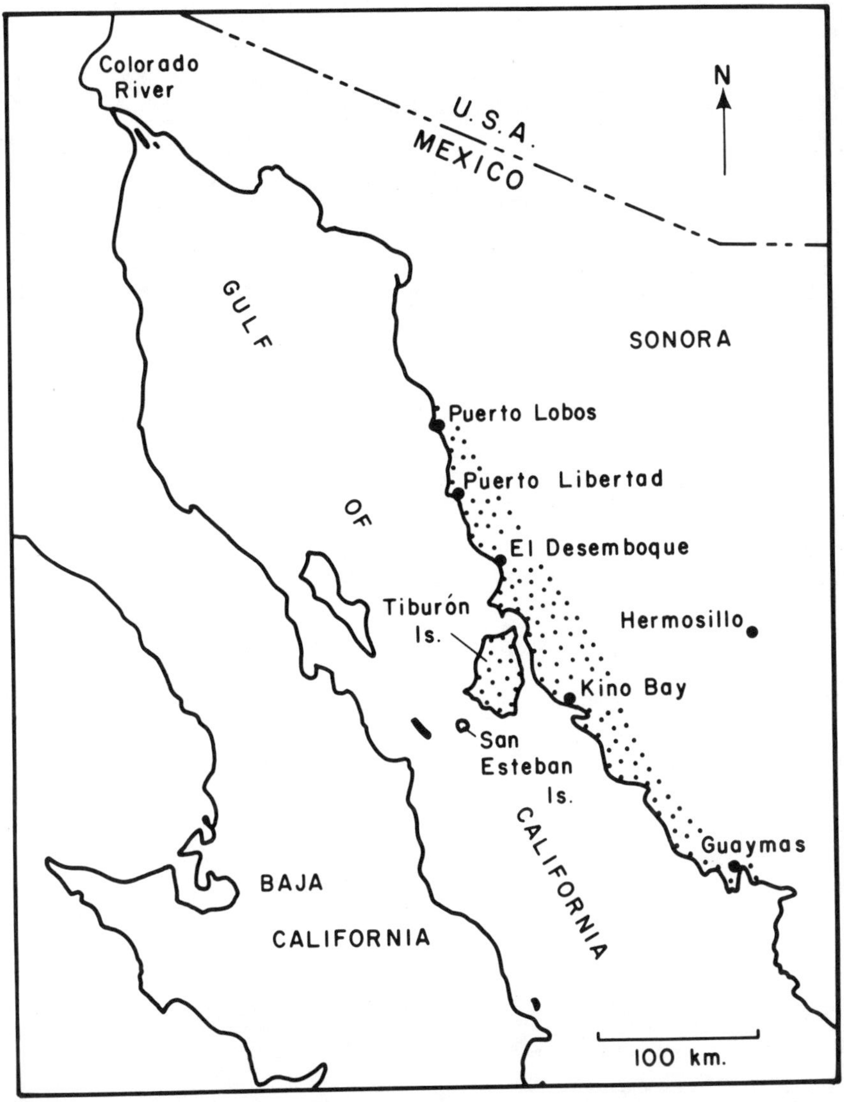

Fig. 2. Northern part of the Gulf of California showing Shreve's Gulf Coast of Sonora vegetational subdivision of the Sonoran Desert. Pre-contact Seri occupation extended along the coast from the vicinity of Guaymas in the south to Puerto Lobos in the north and included Tiburon and San Esteban Islands.

Table 1. Relationships of numbers of species of food plants to area. Species numbers are for seed plants, and food plants are those palatable to man.

	San Esteban Island	Tiburon Island	Entire Seri Region	Sonoran Desert	World
Total flora	108	290	500	2,500	200,000
Food plant species	23	60	80	375*	30,000 ?
Major food species	6	8	10	38-40	3,000 ?
Area (km sq.)	43	1,200	22,000	310,000	147,200,000**

*Native species **Land surface

An urgent task is that of reclaiming and restoring usefulness to human-disturbed ecosystems. For example, thousands of hectares of farmland in southern Arizona have fallen out of alfalfa and cotton cultivation within this decade, due to plummeting groundwater levels and skyrocketing pumping costs. The abandonment of these fields has left large areas vulnerable to wind erosion creating a mini Dust Bowl. Seeding abandoned fields with native species which can form protective ground cover and offer harvestable byproducts is now a viable alternative. There are certain root perennials, such as the buffalo gourd, which readily colonize disturbed soils, and that could provide fruit, root or oilseed crops with little or no irrigation, since they are pre-adapted to aridity.

Growth-form diversity may also allow mixed cropping of economic plants without creating problems in management or harvesting. The tepary bean *(Phaseolus acutifolius)*, an ephemeral, could be grown and harvested in the interspaces between perennial shrubs such as jojoba *(Simmondsia)* or mesquite *(Prosopis)* before they are large enough to produce economic harvests (Nabhan and Felger in press). Flannery (1973) has described the twining of wild beans and squash vines on maize's closest relative, teosinte. Just as this natural relationship may have been emulated by early agriculturalists, the multiple strata of natural ecosystems can serve as a model in designing new agroecosystems (Wiseman in press).

The harvest of wild crops will not alleviate present-day and near-future shortages of food, energy, and fresh water. Rather, we advocate that wild and formerly cultivated crops such as those briefly mentioned here be developed for agricultural exploitation by the implementation of hybridization, selection, and other means of genetic improvement and associated research. It is often pointed out that people do not readily accept new kinds of food, and food preferences are another important anthro-agronomic concern. We suggest that new foods should be utilized within the cultures developing them before encouraging other cultures to accept them. While this may be overly simplified, diffusion of behavior traits from economically dominant cultures is readily observable in the present-day world. Furthermore, rapid diffusion and acceptance of new foods has certainly recurred throughout history, such as in the aftermath of what Crosby has called the "Columbian exchange" (Crosby 1972).

Many factors enter into the selection of new crops, and these will vary according to the desired product, growth-form, and microhabitat to be exploited (see discussion by Evans 1975 on new ideotypes). Critical characteristics include productivity, nutritional quality, hybridization potential, pest resistance; ease of harvesting, storage and preparation; and water, fertilizer, and tillage requirements. Many major food species native to the Southwest should score high in this regard. For example, Felker (1976) provides evidence that mesquite *(Prosopis)*, in the legume family, should be an ideal arid-land food crop (see discussion below on desert trees). Among certain legumes nitrogen-fixation is certainly an advantageous character. Present data indicate that nutritional quality and productivity are indeed high among many of the major food plants from the Sonoran Desert region (see Jones and Earle 1966, Earle and Jones 1962, Cruz 1973, Felger 1975, Felger and Nabhan 1976, Felker 1976, and National Academy of Sciences 1976).

Pre-contact subsistence patterns in the Sonoran Desert region consisted of 1) agriculture combined with gathering and hunting in the better watered areas and 2) hunting and gathering without agriculture in the more arid areas (Sauer 1935). The major pre-contact cultivated crops in the American Southwest, such as the ancient American trinity of maize-bean-squash, are summer-fall crops attuned to the hot season monsoon of Mexico. While various wild food plant resources were available for harvest at different times of the year, the majority of the more important subsistence species consist of perennials with crops ripening at the end of the long pre-monsoon dry season in late spring and early summer.

Disruption and eventual discontinuation of the harvesting of many wild food resources began with Spanish colonial missionization and introduction of Old World crops. The major European introductions, such as wheat, are winter crops in the American Southwest, harvested in late spring. This provided at least a partial alternative for earlier dependence on pre-monsoon dry season wild crops such as mesquite *(Prosopis spp.)*. The *reducción* system, based upon European preconceptions of proper human organization, in which dispersed Indian populations were concentrated around the missions, tended to discourage harvesting of the relatively dispersed wild resources (Spicer 1962). Although indigenous knowledge and practices diminished greatly during the centuries of post-contact acculturation, a surprising quantity of information persisted well into the 20th century. Native knowledge and practices are presently relictual and disappearing faster than they are being recorded.

The demise of interest in indigenous crops, both cultivated and wild, occurred primarily during times of plenty, when there appeared to be no serious shortages of food, inexpensive energy, and fresh water, and when population levels, both local and worldwide, were substantially lower. Projected future food, energy, and water shortages all indicate a re-evaluation of energy-intensive agriculture with high water requirements, particularly in dry lands.

To illustrate, some specific examples from the Sonoran Desert are briefly discussed below. We are investigating several species within each of six major growth-forms within the Sonoran Desert: ephemeral herbs, columnar cacti, desert trees and shrubs, root perennials, seagrasses, and saltgrasses.

Ephemerals

As annual plants, ephemerals complete their life-cycles in a single season or less, and are characteristic of the hot deserts of the world. Ephemerals pass through times of drought by remaining in the soil as seed, thus avoiding stress. These plants have evolved to exploit the short, sporadic periods when conditions become non-desert like, i.e., when soil moisture is high. Growing quickly once they emerge, desert ephemerals seem to put a greater percentage of energy directly into reproductive effort (seed production and quality) than most growth-forms. Their ratio of seed yield to vegetative productivity is therefore unusually high.

In the Sonoran Desert there are two major sets of ephemerals adapted to the biseasonal rainfall of the region: winter-spring or cool season species, and summer-fall or hot season species (Shreve 1951; Mulroy and Rundel 1977). Winter-growing species remain dormant through the hot weather rainy period even with sufficient soil moisture, and vice-versa with summer ephemerals. In addition, some species of ephemerals are non-seasonal and respond to soil moisture at any time of the year. Each season brings a different species spectrum and several sets are possible each year. Often an entire ephemeral population will emerge simultaneously to form an extensive, single species stand. Some can complete their life cycle in only four to six weeks, while others require several months. Because of their short life cycles, ephemerals are especially susceptible to genetic improvement and might be developed into new agronomic crops within a relatively few years.

Some desert ephemerals have hybridization potential with congenerics which have long been in cultivation in various parts of the world, e.g., *Amaranthus*, *Phaseolus*, and *Chenopodium*. For the most part desert ephemerals have evolved from non-desert perennial ancestors (Shreve 1951). Similarly most of the major world crops, the economic annuals, have been developed from perennial ancestors (Ames 1939). Further "ephemeralization", or shorter life cycles, should be possible.

Lesquerella fendleri (Cruciferae or Brassicaceae), a wild mustard from the Chihuahuan Desert region, has spontaneously produced stands large enough and with sufficient productivity to be harvested by combine (Gentry and Barclay 1962). There are closely related species in the Sonoran Desert and hybridization potential is high (Rollins and Shaw 1973). The tiny seeds contain 20-28% oil and 20-22.5% protein (Jones and Earle 1966). The seeds and to some extent the greens of other ephemeral mustards - such as *Descurrainia pinnata*, tansy mustard, and *Lepidium lasiocarpum*, a peppergrass - were important foods for early Native Americans (Bye 1972, Castetter and Bell 1941, 1951). The seeds are high in protein and oil content (*D. pinnata*, 24.4% protein and 38.4% oil; *L. lasiocarpum*, 26.3% protein and 23.5% oil; Earle and Jones 1966).

Ethnohistorically, wild and domesticated forms of the tepary bean, *Phaseolus acutifolius* (Leguminosae or Fabaceae) have long been significant in southwestern North America. The cultivated, domesticated tepary has been a food resource in the Americas for more than five millenia (Kaplan 1965). The value of this very productive and protein-rich legume is enhanced by the fact that it does well even in fairly arid environments. Domesticated teparies can produce as much as 2,000 kg/seed/hectare in water supplemented fields within the Sonoran Desert and more than 4600 kg/seed/hectare in areas where climatic conditions are less extreme (Hendry 1918, 1919). The beans store well, and are nutritionally comparable or superior to most cultivated legumes, with protein content ranging from 21.2% to 32.9% (Nabhan and Felger in press).

Use of wild teparies appears to have been discontinued. Recent subsidence of tepary cultivation is related to the breakdown of traditional economies and land use, and to the introduction of energy-intensive irrigation agriculture. However, certain domesticated varieties are still grown by local commercial and subsistence farmers. An unsuccessful attempt earlier in the century to introduce teparies into modern agriculture was poorly timed. Nevertheless, the tepary has considerable potential for low maintenance agriculture in dry lands and may become an important crop (Nabhan and Felger in press).

Seeds and in some cases the herbage (greens) of numerous other ephemerals featured prominently in indigenous diets. Among them were species of *Amaranthus* (Amaranthaceae), *Chenopodium* (Chenopodiaceae), *Oligomeris* (Residaceae), *Mentzelia* (Loasaceae), and *Proboscidea* (Martyniaceae). The devil's claw, *Proboscidea parviflora,* is particularly interesting. The seeds are relatively large (about 6 X 10 mm) with a protein content of 22.5% to 25.5%, and a high-quality oil content of 36% to 40% (James W. Berry, personal communication). The wild type, cultivated under simulated dryland conditions, has yielded 503 kg of oil from 1120 kg of seed/hectare (Paur 1952). A domesticated variety, much more suited to agronomic conditions, is already being grown as a cash crop in the Papago Indian basketry fiber industry (Nabhan and Fritz 1977).

Root perennials

These plants have large underground storage organs which send up vegetative shoots for a short period of time during favorable conditions. They are similar to ephemerals in that they rapidly grow and produce seed under favorable, moist conditions, but otherwise remain dormant. The enlarged roots are generally high in carbohydrate content, the seeds may be rich in protein or oil, and some have other useful parts. However, there are relatively few species of agronomically interesting root perennials in the Sonoran Desert.

Phaseolus metcalfei, corcomeca (Leguminosae), produces edible beans, while its taproot is a traditional medicine and fermenting agent (Castetter and Bell 1942, Gentry 1942, 1963). Every part of *saiya, Amoreuxia palmatifida* (Cochlespermaceae) is edible, although only the parsnip-like root has been a marketed commodity (Felger 1975, Gentry 1963). The buffalo gourd, *Cucurbita foetidissima* (Cucurbitaceae), provided oil-rich seed to indigenous peoples for centuries, and has an enormous root potentially useful as an industrial starch. It and other xerophytic cucurbits are currently being tested for agronomic adaptability by an international team (Bemis 1975).

Columnar cacti

These large cereoid cacti (*Cactaceae*, subfamily Cactoidae) have one or more erect stems and attain shrub to tree heights. The larger ones weigh many tons. Within this group there is a considerable span of ecological tolerances. Some are extremely drought resistant, such as *cardón*, *Pachycereus pringlei*, occurring with an average annual rainfall of 100 to 200 mm and extreme heat. It grows on steep slopes as well as level terrain and tolerates moderately saline soils.

Columnar cacti are long-lived and high yields can be expected over many decades. Fruiting can occur in about a decade from seed, about the time required for most tree crops. Others, such as the saguaro, *Carnegiea gigantea*, may require 30 to 60 years. However, reproduction might be facilitated with propagation by cuttings, grafting, and tissue culture (Mauseth 1977).

Columnar cacti provided important food resources and wines for New World peoples (Castetter and Bell 1937, Felger and Moser 1974). Fruit yields are often substantial, with considerable variation in productivity, fruiting time, taste, size, and nutritional quality among the diverse species. Fresh weight for fruit of several Sonoran Desert species is 150g/fruit or more, and 50 to 100% of the fruit is edible. Typically there is a sweet, sugar-rich juicy pulp in which are embedded numerous small protein and oil-rich seeds. For example, seeds of *cardón* contain 22.6% protein and 32.2% edible oil (Felger and Moser 1976). The quality and quantity of sugar content of the pulp is exceedingly high (Greene 1936, Pant 1973), and it might be suitable for wineries and non-alcoholic beverage industries in arid tropical regions.

On the order of 100 to 200 species of columnar cacti in arid and semi-arid regions of North and South America and one half dozen species in the Sonoran Desert region merit agronomic investigation. This great diversity would allow various combinations of characters to fit different environmental and agronomic needs. Broad, intergeneric hybridization is possible among many cacti (Friedrich 1974); however, hybridization potential among the columnar cacti remains largely untested.

Desert trees and shrubs

Among the most important harvests of pre-Columbian peoples in the Sonoran Desert region were the fruit and seeds of certain common trees and shrubs. At higher elevations adjacent to the desert, the abundant crops of various trees such as the Emory oak or *bellota, Quercus emoryi* (Fagaceae), and piñon, *Pinus edulis* (Pinaceae), continue to be harvested. Within the Sonoran Desert itself approximately ten arborescent legume species were of major significance to early Native Americans, e.g., *Prosopis spp., Cercidium spp., Olneya tesota,* and *Lysiloma spp.* A wide range of characters of potential agronomic importance are variously present among these species. The seeds, relatively large in some species, are high in protein content, and some have indehiscent pods with carbohydrate-rich mesocarp ("pulp"). Indehiscent pods are agronomically desirable because the contents are not lost in harvest or storage. None of these tree crops require tillage. Many legume trees are nodulated with nitrogen fixation rates of 200 ± 45 kg N/ha/year (Orchard and Darb 1956, Allen and Allen 1959, Felker 1977).

At present the several dozen species of mesquite *(Prosopis:* section *Algarobia)* in both North and South America offer the broadest spectrum of characters of potential agronomic value (Felker 1976). When fully ripe and sweetest, the indehiscent pods fall to the ground facilitating harvest. As with most other desert legume tree crops, ripening of the fruit on all trees in each area occurs at nearly the same time, and in some regions two crops per year may be possible. There is considerable variation in size, taste, and yield of pods of mesquite as well as other major food resource species, and different groups annually harvested specific, superior trees and groves (Felger in press, Felger and Moser 1971).

The mesocarp or "pulp" of the pod provided a major carbohydrate or calorie-rich component in native diets. Mesquite appears to be relatively independent of short-term drought, primarily due to its deep roots which commonly penetrate to 20 m to reach ground water (Solbrig and Cantino 1975). The deepest known roots in the world are those of a mesquite near Tucson recovered from a mine at a depth of 53 m (175 feet; Phillips 1963).

Approximately one half of the total weight of mesquite pods consists of sweet mesocarp tissue (Walton 1923) which can be made into flour, mixed with water, and consumed as an unleavened cake or bread, or as a beverage. No cooking is required, which would be significant in fuel-starved regions of the world. The protein-rich seed comprises about 15% of the total pod by weight, and the seed kernel (seed with the seed coat removed) 8% of the pod (Walton

1923). The kernel contains 69% protein, which exceeds that of the soybean (Felker 1976). Difficulty in separating the true seed from the encasing, inedible stony endocarp or pit may limit use of the seed. Yet some ancient desert peoples devised stone gyratory crushers which can still be utilized to break open the endocarp and free the seed as well as process mesocarp flour. Modern gyratory crushers could be devised to do the same jobs more efficiently (Felger and Nabhan 1976).

Productivity of wild mesquite and several other major Sonoran Desert legume trees and shrubs is high: our data indicate yields in excess of 1,000 kg/pods/hectare when extrapolated to single species stands. Under agricultural conditions and with genetic improvement productivity should be still higher.

Grazing pressure and seed dissemination by livestock have contributed to the spread of mesquite, and major research efforts are presently being directed toward its eradication as a pest species. However, in earlier millenia mesquite was the single most widespread and valuable resource among the diverse peoples of arid and semi-arid regions of southwestern North America. It was utilized for food, fuel, fiber, lumber, tools, dye, medicine, and many other practical and aesthetic purposes (Bell and Castetter 1937, Doelle 1977, Felger and Moser 1971, Felger in press).

Seagrasses

Eelgrass, *Zostera marina* (Zosteraceae), grows submerged in shallow seas along the coast of much of the Northern Hemisphere. In contrast to other known populations, in the Gulf of California 100% of the stems produce seeds. Our data indicate that seed productivity can be 2400 kg/ha/year. The Seri Indians harvested seed from the great masses of eelgrass which drift ashore each spring, and prepared it into flour which served them as a major food resource (Felger and Moser 1973, Felger and McRoy 1975, Felger *et al.* in press).

A seed crop which can be grown in the sea, and presumably requiring little or no fertilizer, pesticide, or tillage, would be of great value. However, agronomic development of a crop submerged in tidally active water offers many unsolved problems and may be a costly enterprise. Hybridization has not been attempted for seagrasses, although there may be some such potential. There are about one half dozen species of *Zostera* closely related to *Z. marina*. *Z. marina* seed is three mm long and that of

Z. asiatica in the northeast Orient is five to six mm long.

Surfgrass, *Phylospadix* (Zosteraceae), on the Pacific coast of North America and Japan, with two closely related species, should also be investigated for adaptation to sea farming. Seed productivity might be on the order of that of eelgrass. Surfgrass roots form a tenacious holdfast, anchoring the plant to rocks in open surf. Perhaps it could be grown on racks floated at sea on rafts, and moved about to obtain favorable microenvironmental conditions.

Saltgrass

Tidal flats at the delta of the Colorado River once supported many thousands of hectares of Palmer saltgrass, *Distichlis palmeri*. This coarse perennial grass thrives with brackish or sea water, developing a dense growth which is highly drought resistant. Because of this, competition with weeds should be minimal. The grain is 6.5 mm long (about as large as wheat), and there should be considerable potential for hybridization.

The Cocopa harvested the grain in May as a major staple. They ground it into a coarse flour, usually eating it as a gruel. It may also be made into leavened or unleavened bread (Vasey 1889, Castetter and Bell 1951). Hardy (1829) reported it as "...wild wheat, the taste of which is very sweet." While the nutritional content of the grain remains unknown, it should be high.

The spikelets readily shatter when harvested and are easily dislodged by tidal action. Grains that shatter are usually undesirable because of difficulty in recovering the grain from the ground. However, in this case the spikelets float, and are deposited in tidal windrows. It might be possible to devise methods of skimming floating grain from the water or harvesting it from windrows. Significantly, the grain seems to separate readily from the chaff. It can also be harvested while still green and then dried, which would overcome the problem of early shattering (Felger 1975).

Summary

We suggest that substantial increases in agricultural diversity can be achieved by increasing both the growth-forms and the number of species of agricultural crops.

Increases in human population are associated with general decreases in the number of subsistence species -

including loss of agricultural diversity. Presently only a handful of species support civilization. Large energy and fresh water inputs are needed to simulate tropical and temperate agroecosystems needed by industrial agriculture in dry lands. In contrast, indigenous Sonoran Desert peoples utilized about 375 native species of wild food plants, or about 15% of the flora. Extrapolated worldwide, this allows a crude estimate of 30,000 species of food plants with 3,000 major subsistence species. Certainly some of these should be suitable for agronomic development to substantially increase agricultural diversity.

The most important world food crops are annuals with relatively little diversity in their environmental requirements. In contrast, major wild subsistence species include a broad spectrum of growth-forms adapted to highly diverse ecosystems. In the Sonoran Desert these include seagrasses, saltgrasses, desert trees and shrubs, columnar cacti, root perennials, and desert annuals (ephemerals). Many of these are excellent candidates for agronomic development in the Southwest because of high productivity, high nutritional values, and their adaptation to local conditions. Possibilities include grain crops which can be grown with saline water, or even sea water, and dry land crops requiring little or no irrigation.

We believe that there are important indigenous resource species in each part of the world suitable for agronomic development. Future agroecosystems can become ecologically well adapted to local conditions if scientists and planners from various fields can collaborate, so as to overcome the inertia and/or prejudices associated with the cultivation and use of unfamiliar plants.

Acknowledgements

We thank the following persons for encouragement and information: William P. Bemis, James W. Berry, Dennis O. Cornejo, Peter J. Felker, Bernard L. Fontana, Howard Scott Gentry, Richard Levins, Paul S. Martin, Jean Mayer, Mary Beck Moser, Barbara L. Tapper, and the members of the Tepary Burrito Society of Tucson. We are grateful to Barbara Tapper for preparation of the manuscript. Supported by National Science Foundation (SOC 75-13-628).

References

Allen, Ethel K., and O.N. Allen
1961 The Scope of Nodulation in the Leguminosae. Recent Advances in Botany, IX International Botanical Congress, Montreal, 1959. Vol. I. Pp. 585-588. Toronto: University of Toronto Press.

Ames, Oaks
1939 Economic Annuals and Human Cultures. Cambridge: Botanical Museum of Harvard University.

Bell, Willis H., and Edward F. Castetter
1937 The Utilization of Mesquite and Screwbean by the Aborigines in the American Southwest. University of New Mexico Bulletin 314, Biological Series 5(2):1-55.

Bemis, W. P., et al.
1975 The Buffalo Gourd *(Cucurbita foetidissima)* HBK: A Potential Crop for the Production of Protein, Oil and Starch on Arid Lands. Office of Agriculture, A.I.D, Technical Series Bulletin No. 15, Washington, D.C.

Bohrer, Vorsila L.
1975 Plants That Have Become Extinct in the Southwest. Symposium on Threatened and Endangered Plants. New Mexico Sections of the Society for Range Management and Society of American Foresters. Unnumbered manuscript, Albuquerque.

Bye, Robert Jr.
1972 Ethnobotany of the Southern Paiute Indians in the 1870's. *In* Great Basin Cultural Ecology, A Symposium. D. D. Fowler, ed. Pp. 87-104. Desert Research Institute Publication in the Social Sciences no. 8.

Castetter, Edward F., and Willis H. Bell
1937 The Aboriginal Utilization of the Tall Cacti in the American Southwest. University of New Mexico Bulletin 307, Biological Series 5(1):1-48.
1942 Pima and Papago Indian Agriculture. Albuquerque: University of New Mexico Press.
1951 Yuman Indian Agriculture. Primitive subsistence on the Lower Colorado and Gila Rivers. Albuquerque: University of New Mexico Press.

Crosby, A. W. Jr.
1972 The Columbian Exhange: Biological and Cultural Consequences of 1492. Contributions in American Studies No. 2. Westport, Connecticut: Greenwood Press.

Cruz, Robert
1973 Desert Plant Chemurgy: A Current Review. Economic Botany 27:210-230.

Earle, F.R., and Quentin Jones
1962 Analyses of Seed Samples from 113 Plant Families. Economic Botany 16:221-250.

Evans, L. T.
1975 The Physiological Basis of Crop Yield. *In* Crop Physiology. L. T. Evans, ed. Pp. 327-355. London: Cambridge University Press.

Felger, Richard S.
1975 Nutritionally Significant New Crops for Arid Lands: A Model from the Sonoran Desert. *In* Priorities in Child Nutrition in Developing Countries. Jean Mayer and J. W. Dwyer, eds. Pp. 373-403. New York: UNICEF.
in press Mesquite in Southwestern Indian Cultures. *In* A Tree in Perspective: *Prosopis* in Desert Shrub Ecosystems. B. Simpson, ed. Stroudsberg, Pa.: Dowden, Hutchinson and Ross.

Felger, R. S. and C. H. Lowe
1976 The Island and Coastal Vegetation and Flora of the Northern Gulf of California, Mexico. Natural History Museum of Los Angeles County, Contributions in Science No. 285.

Felger, R. S., and C. P. McRoy
1975 Seagrasses as potential food plants. *In* Seedbearing Halophytes as Food Plants. G. F. Somers, ed. Pp. 62-74. Newark: University of Delaware.

Felger, R. S., and M. B. Moser
1971 Seri Use of Mesquite *(Prosopis glandulosa var torreyana)*. The Kiva 37:53-60.
1973 Eelgrass *(Zostera marina* L.) in the Gulf of California: Discovery of its Nutritional Value by the Seri Indians. Science 181:355-356.
1974 Columnar Cacti in Seri Indian Culture. The Kiva 39:257-275.
1976 Seri Indian Food Plants: Desert Subsistence Without Agriculture. Ecology of Food and Nutrition 5:13-27.

Felger, R. S., M. B. Moser, and E. W. Moser
in press Seagrasses in Seri Indian Culture. *In* Studies of the Seagrass Ecosystem. R. Phillips and C. P. McRoy, eds. New York: Marcel Dekker.

Felger, R. S., and Gary P. Nabhan
1976 Deceptive Barrenness. Ceres 9(2):34-39.

Felker, Peter J.
1976 Potential Utilization of Leguminous Trees for Minimal Energy Input Agriculture and Amino Acid Analysis by Gas-Liquid Chromatography. PhD. dissertation, Department of Botany and Plant Pathology, Michigan State University.

Flannery, Kent V.
1973 The Origins of Agriculture. Annual Review of Anthropology, Vol. 2. Pp. 271-310.

Friedrich, Heimo
1974 Hybridization as a Factor in the Evolution of the Cactaceae. Cactus and Succulent Journal 46:213-214.

Gentry, Howard Scott
1942 Rio Mayo Plants. A Study of the Flora and Vegetation of the Valley of the Rio Mayo, Sonora. Carnegie Institution, Washington, Publ. 527.
1963 The Warihio Indians of Sonora-Chihuahua: an Ethnographic Survey. Bureau of American Ethnology Bulletin 186:61-144.

Gentry, H.S., and A. S. Barclay
1962 The Search for New Industrial Crops III: Prospectus of *Lesquerella Fendleri*. Economic Botany 16:206-211.

Greene, R. A.
1936 The Composition and Uses of the Fruit of the Giant Cactus *(Carnegiea gigantea)* and its Products. Journal of Chemical Education 13:309-312.

Hardy, R. W. H.
1829 Travels in the Interior of Mexico in 1825, 1826, 1827 and 1828. London: Colburn and Bentley.

Hendry, G. W.
1918 Bean Culture in California. University of California Agricultural Experiment Station Bulletin 294.
1919 Climatic adaptation of the white tepary bean. Journal of the American Society of Agronomy 11:247-252.

Jones, Quentin, and F. R. Earle
1966 Chemical Analyses of Seeds II: Oil and Protein Content of 759 Species. Economic Botany 20:127-155.

Kaplan, Lawrence
1965 Archaeology and Domestication in American *Phaseolus* (Beans). Economic Botany 19:358-368.

Mauseth, James D.
1977 Cactus Tissue Culture: A Potential Method of Propoagation. Cactus and Succulent Journal 49:80-81.

Mulroy, T.W., and P. W. Rundel
1977 Annual Plants: Adaptations to Desert Environments. BioScience 27:109-114.

Nabhan, Gary, and R. S. Felger
in press Biogeography and Ethnohistory of Teparies *(Phaseolus acutifolius)* in Southwestern North America. Economic Botany.

Nabhan, Gary, and Gordon Fritz.
1977 Devil's Claw *(Proboscidea)*: Cash Crop in the Papago Basketry Industry. Journal of the Arizona Academy of Science 12:4-5.

National Academy of Sciences
1975 Underexploited Tropical Plants With Promising Economic Value. Washington, D.C.: National Academy of Sciences.

Orchard, E.R., and G. D. Darb
1956 Fertility Changes Under Continued Wattle Culture With Special Reference to Nitrogen Fixation and Base Status of the Soil. Proceedings of the 6th International Soil Science Congress, Paris. Vol. IV:305-310.

Pant, Radha
1973 Chemical Analysis of *Cereus sp.* Cactus Fruits. Current Science 42:721-724.

Paur, Sherman
1952 Four native New Mexico Plants of Promise as Oil Seed Crops. Press Bulletin 1064, New Mexico Agricultural Experiment Station. State College, New Mexico.

Phillips, W.S.
1963 Depth of Roots in Soil. Ecology 44:424.

Phillips, S.H., and H. M. Young Jr.
1973 No Tillage Farming. Milwaukee: Rieman Associates.

Rollins, R. C., and E. A. Shaw
1973 The Genus *Lesquerella* (Cruciferae) in North America. Cambridge: Harvard University Press.

Sauer, Carl O.
1935 The Aboriginal Population of Northwestern Mexico. Ibero-Americana 10.

Shreve, Forrest
1951 Vegetation of the Sonoran Desert. *In* Vegetation and Flora of the Sonoran Desert. Forrest Shreve and Ira L. Wiggins, eds. Part 1. Carnegie Institute, Washington, Publ. 591.

Shreve, Forrest, and Ira L. Wiggins
1964 Vegetation and Flora of the Sonoran Desert. 2 vols. Stanford: Stanford University Press.

Solbrig, O. T., and P. D. Cantino
1975 Reproductive Adaptations in *Prosopis* (Leguminosae, Mimosoideae). Journal Arnold Arboretum 56:185-210.

Spicer, E. H.
1962 Cycles of Conquest. Tucson: University of Arizona Press.

Vasey, George
1889 New or Little Known Plants. Garden and Forest 2:401-402.

Walton, G. P.
1923 Nutritional Value of the Beans of *Prosopis juliflora, Ceratonia siliqua and Gleiditschia triacanthos*. United States Department of Agriculture, Bulletin 1194.

Wiseman, Frederick
in press. Agricultural and Historical Ecology of the Maya Lowlands. *In* Prehispanic Maya Agriculture. P. D. Harrison and B. L. Turner, eds. Albuquerque: University of New Mexico Press.

In this, the final paper, John Bennett presents a new synthetic approach, based on rational choice and general systems theories, for thinking about and predicting the farmer's behavior in different agricultural resource management schemes. Though his examples come largely from the United states and the Canadian plains, he makes it clear that his model should also work for more traditional horticultural systems, as well as for areas where agrarian development is occurring. He includes not only water, land, and crops, but also social organization and ideology in the model. The paper contrasts with and complements the others in that it is basically an attempt at theory-building. Eventually, of course, we must provide useful models, and not just isolated examples, if we are to be of real use to policy makers. Bennett's paper ties together some of the otherwise loose ends in this collection.

6

A Rational-Choice Model of Agricultural Resource Utilization and Conservation

John W. Bennett

Introduction

This paper describes a way of thinking about cultural components of the utilization of agricultural or "renewable" resources[1]. By 'model' I refer to a means of ordering or constraining thought, and not a rigorous mathematical scheme. The approach described is a freehand application of a particular theoretical development in behavioral science, and one of my objectives is to suggest a way to make that field more responsive to the need for a behavioral theory of agricultural resource management. While the model pertains to agricultural production, especially in environments with severe resource constraints, applications to other contexts may be evident.

Sociologist Walter Firey pointed out in a pioneering 1959 work that several schemes were available for the analysis of resource utilization. He described three: The ecosystemic concerned disturbances in natural systems created by extraction of substances and organisms; the second, the ethnological, pertained to culturally-styled means of using resources characterizing particular groups; and the third, the economic, was described as an analysis based on concepts of scarcity and gain. Firey also contributed his own approach, an elaborate classificatory

The author is indebted to Dr. Robert Klepper, of the Center for the Biology of Natural Systems, Washington University, for clarification of a number of key points from the viewpoint of an economist.

scheme featuring the concept of "resource complex", or organized set of behaviors dealing with physical substances.

Firey's book appeared before two important developments had a chance to make their mark: rational choice and/or social exchange theory; and systems theory.[1] The former emerged in the behavioral sciences on the basis of economic models, the latter in several scientific fields more or less simultaneously as an offshoot of the cybernetics movement and the developing field of bio-ecology. As a model for the investigation of resource utilization, general systems theory has obvious advantages over ecosystem theory since the former is not biased toward biology, and is technically capable of handling inputs from all possible sources. Theories of rational choice and social exchange were just beginning to appear when Firey produced his book, and I shall have more to say about them in a moment.

In the background of this paper are certain problems of handling human behavioral data in general systems approaches to resource utilization and conservation. The systems approach takes the point of view of the omniscient scientist who views physical and social components of resource practices as variables of equal strength and as subject to analysis with equal objectivity. Human choice is seen mechanistically: as an energy charge or a fork in a decision tree. Resource management is viewed as a set of rules governing the operation of feedbacks in the system. At the other extreme, cultural anthropologists view resource practices as human behavior occurring in unique, substantive configurations of social and environmental factors.[2] Symbolic, behavioral, technological or mechanical, and environmental components of the resource utilization system are sharply distinguished, in contrast to the general systems approach, which tends to treat them as if they were all alike.

While the particularism of the anthropological approach makes generalization difficult--and this is <u>its</u> chief difficulty--it is possible to define its basic goals and assumptions. The first and most fundamental is that human systems are adaptive,[3] which means that they are dynamic (subject to unpredictable change); are frequently governed by factors which originate outside of the system under analysis (e.g., prestige valuations in consumption behavior); and are more unstable and changing than most natural systems. The element of time is also crucial; human adaptive systems can change more rapidly and at rates which vary less coherently than is the case for many natural systems.[4]

A second characteristic feature of human systems is a chain of problems and solutions, each solution begetting another problem, which in turn must be solved, and so on. At each point in the cycle or chain, new factors can enter the equations and the forces exerting control may require drastic overhaul if some degree of continuity or predictability is to be maintained. The process just described is called behavioral or social adaptation, and while its structure is not necessarily peculiar to humans, certainly it is quantitatively more significant in the human species than in any other.

Third, cognitive functions govern or influence this process. The distinctive feature of human cognition is its time-binding or anticipatory characteristic: this means that purposiveness or goal-orientation is more complex in human behavior than it is in other species. More precisely, purposive behavior in humans is reinforced and elaborated by the ability to conceive of time as a commodity and hence anticipate outcomes, or govern present actions in accordance with a time scale. As economists put it, humans maximize _expected_ utility: the gain as qualified by its probability of occurrence--which is a temporal function. These processes are also fundamental in what we call social and economic growth; that is, the tendency to accumulate stocks of material, commodities, and symbols called "capital", which permit transformations of physical resources to produce more such values, which in turn can be utilized as capital, etc.

Although the historical record is irregular, the overall tendency in human consumption of resources since the appearance of _Homo sapiens_ is roughly exponential with key upswings at the basic resource-transformation points of plant and animal husbandry (_ca_. 7000 B.C.); urbanized trade and commerce (_ca_. 3000 B.C.) or powered machinery (_ca_. 1700 A.D.).[5] The term "conservation" donates the need to control, stop or reduce something; hence carries the admission that human behavior is likely to exceed prudent limits with respect to the use of physical resources. This implies further that there is a bias or tilt in human behavior which we are now having to cope with on a world-wide scale. One essential problem of control is to find ways to bring our use of renewable resources under sustained-yield management; that is, to remove from the production process those motives and incentives which derive from institutions and forces outside the ecological or resource-budgeting nexus.

Another way of viewing the problem is represented by Garrett Hardin's concept of the "tragedy of the commons" (Hardin 1968). In this concept, the anticipatory-growth-accumulative tendencies in human behavior are dramatized as the tendency for indivudal users to increase the use of collective-use facilities like grazing lands until the quality, quantity, or productivity of the lands in question are reduced for everyone concerned. From a sociological standpoint, this conceptualization concerns the equity drive in human societies; it is a distinctive social manifestation of the generalized anticipatory function, and requires the presence of numbers of persons who compete with each other for access to resources, or who insist on equal shares for various reasons--including the most fundamental one which is that someone else has acquired a share.

For whatever reason, the behavioral tendency toward accumulative overuse with respect to resource use is now an established fact. No one knows for certain, but there is evidence that deserts in Africa and the Middle East and possibly the Americas are increasing due to human pressure on friable resources, caused by increasing human population, migration, and the application of particular technologies of food production.[6] This is the case even for "low-energy" farmers, and tribal pastoralists, whose resource-use systems were mainly benign under smaller magnitudes of populations of humans and animals. This was probably correct, since tribal peoples in general have been more inclined to maintain balanced ratios between key factors in the ecological nexus than the urbanized and industrial societies, although there are also more cases of tribal resource abuse than is generally believed by conservationalists who tend to idealize "primitive man" and his ecological posture.[7] That is, while the thrust of *Homo sapiens*' history is toward exponential use of resources and toward overuse or abuse, this history contains segments and populations in which the growth emphasis is less evident than in others. However, this fact should not be allowed to give rise to optimistic predictions of our ability to develop resource-use control systems which avoid the need for strong sanctions.[8]

Rational Choice Models and Agricultural Decisions

The task is, then, to think about resource use and conservation with reference to key characteristics of human behavior, or to view technological and economic factors as forms of behavior. Such a model should incorporate factors which stimulate resource use and abuse; i.e., the peculiar temporal orientation or "future-binding" of behavior; the proclivity toward accumulative or growth-oriented activity; and the ability of humans to make choices and decisions on the basis of these factors.

A source of such a model is found in rational choice and social exchange models developed in the behavioral sciences on the basis of concepts of scarcity and consumer behavior in economics.[9] The basis of this model is the "law of scarcity"; hence people must make choices between things which they need or want. That is, since another key trait of human behavior is the ability to conceive of unlimited wants (there is no reliable, psychological built-in limit, just as there is no reliable built-in institutional control over resource use), scarcities constantly emerge as cognitive factors--not simple physiological urges. Because things from the human viewpoint exist in limited or variable quantities, we must decide on how much of one we will take if we cannot have all we want, or if we need for some reason to make a trade between two things in order to get what we want of one or both.

The development of the rational-choice model by economists probably ranks with other scientific discoveries of the past century, although it should be added that like other key conceptualizations in human behavior its predictive power is modest. In addition, the vast range of phenomena embraced in human behavior include many which are not illuminated by this model. The model is mainly concerned with what is called rational choice: the making of decisions on the basis of thought processes in which qualities and quantities are weighted both subjectively and objectively.

More recently the model was taken up by other social scientists who have attempted to apply its logic and techniques to forms of behavior other than consumption. In sociology, a number of workers following Blau (1964) and Homans (1961) have developed explanations for social transactions of various kinds, including friendships, competition, expertise, wage and hour policies in labor-management relations, and others. For political science, Downs (1957) and Olson (1965), both economists, have attempted an analysis of democracy, power, and collective action based on the rational-choice model. In psychology, Thibaut and Kelly (1959) developed a theory of personality and human interaction based on it. Fredrik Barth (1966) has contributed an anthropological version. Antecedents of the approach are found in the ethnological study of gift exchange and barter economy going back into the 19th century (Mauss 1925).

To some extent, this tradition in social science has provided something comparable to quantitative models of prediction in the natural sciences. It assumes a degree of regularity in human behavior underlying the all-too-evident variability. From a humanistic point of view, this is something of a paradox. Humanists tend to denigrate the social sciences for imposing arbitrary rules on an infinitely plastic and creative human personality. Be that as it may, no species, however spontaneous, has completely random behavior. Even though human behavior has a capacity for novelty and recombination not exceeded by any other species, the manifestations of this novel-response capacity are governed by general properties of the brain and the organism-milieu relationship. The rational choice models have called attention to a behavioral manifestation of this underlying structure - not in language, where most scholars have looked for it, but in the sub-linguistic thought processes vital to human activity.

The social version of rational choice, applied to agricultural production, assumes that choices are made in a social field. The agricultural decision-maker is not an isolate, but is a receiver of signals which come to him from a variety of sources, and with which he has a varitety of personal-social relationships. The nature of these relationships influences his choices and decisions for action. When the relationships are neutral, impersonal, distant--as they tend to be with respect to the "market"--his decisions may be directed solely to the goal of max-mizing benefits to the farm business--a specific behavior classically defined by the economist as "rational". However, in the typical small-scale agricultural enterprise--a world-wide social form--this pure type of decision=making behavior probably rarely occurs, because decisions are made with reference to other relationships, most of them more direct and intimate than those with the impersonal market. Or put another way, the market signals are transmitted by or filtered through other agencies: family members, neighbors, bankers, and other persons with whom the decision-maker has manifold and often long-lasting ties. In this situation, the "rationality" of the decision-making process has to be defined quite differently: it involves finding a way through a maze of often conflicting and hopefully complementing forces and influences. Few farmers have the latitude of choice to permit choosing a crop purely on the basis of profit: they must consider, for example, the burden this particular operation might have on the other family members from the standpoint of labor, or the effects on soil or water of (for example) a heavy use of fertilizer.

Despite the complexity of the decision-making process in a social field, agricultural communities (like other human groups) tend to develop standardized cognitive precedents which function as guides to the interpretation of the economic signals, and solutions to the social-relational issues which arise in the course of choice. These sets of cognitive guides or precedents are often called "a culture" by anthropologists. In any case, and particularly in modern agricultural communities, these cognitive guides can change fairly rapidly, and social relationships likewise, due to the prevailing pragmatic mode of action induced by the economic market system. That is, the definition of rationality created by economists to assist in promoting predictability of outcome has been one of the forces behind culture changes associated with commercialization. Nevertheless, such market-oriented rational choice cannot become total so long as agricultural production is in the hands of private producers involved in communities. Corporate farming may increase the level of rationality, although there are some arguments against this as well.

The rational-choice model thus has two related meanings for the concept of rationality: (1) rational choices can be defined in idealized terms as those which maximize benefits for the firm; or (2) rational choices are those which seek to benefit the firm, but which make compromises between purely economic and other objectives. However, the thought processes involved in both these choice procedures is identical: from a psychological point of view, both are equally "rational" insofar as a procedure of weighing alternatives is taking place. The only difference is that in the second - rational choice conceived as taking place in a social field - a larger number of factors influencing choice are present, and cognitive precedents derived from situations other than "purely economic" are influential in guiding the choice.

Since these cognitive precedents vary greatly from

one social field to another, they are difficult to handle on an aggregate basis--they relate to micro-social situations, not repeated or found in other localities or times, and therefore the market economist really has no way of dealing with them. There is a controversy over the problem among economic anthropologists, and between anthropologists and economists which we shall not discuss here in detail.[10] In essence, the controversy concerns whether the incentives and rules of production in folk or tribal communities are fundamentally different from those in modern communities geared to the impersonal market. The controversy is unreasonable so long as the anthropologist uses a descriptive-integrative method of study, and the economist those quantitative techniques which do not discriminate between cultural and economic phenomena. In my view, the only way this controversy can be resolved is for the economist, who confines research to economic considerations alone - and the anthropologist, who is concerned with economic phenomena in the context of social behavior and the community - to use the same frame of reference: the model of rational choice.[11]

The Model

The model proposed here is concerned with two objectives common to most agricultural systems: yield, or production generally; and conservation, or the maintenance of the basic resources needed for production so that they will continue to provide a yield. These resources can be social, economic, or physical. This paper is concerned with the physical, although it may be evident that the model could be construed to apply to a resource like the supply of farm labor. There are special problems associated with the application of rational choice theory to purely social resources which require a separate discussion, not appropriate here.

The important fact in the relationship between yield and conservation is that they are not in a one-to-one relationship. That is, not all changes in yield produce symmetrical changes in conservation. This is so for two reasons: first, producers can make a variety of choices which contain different trade-offs between yield and conservation; that is, human behavior is not usually restricted to just one choice. Second, technological changes tend to make the relationship flexible or fluid, sometimes permitting higher yields at no altered cost of resource abuse. However, there are limits here, and one of the major problems in world food production is that the ceilings of yeilds at given conservation magnitudes are being reached almost everywhere.

But more fundamentally, any use of soil in crop agriculture, or pasturage in livestock production, probably has some degradational effect, however small. That is, there is no true "renewable" resource anywhere, in all likelihood; it is merely a matter of how much time is required to attain a significant level of abuse or exhaustion. Modern agricultural science has not, to any important degree, adopted this view as its basic term of reference. In its place there has been the growth-oriented assumption that resources are indeed renewable, or at least can be made to yield at desired levels continually or indefinitely. This viewpoint has been the source of serious difficulty in the developed nations, and is becoming a liability in agrarian development programs in arid and semi-arid developing countries.

In the most general sense, conservation of resources can be reached in two ways: (1) by not using certain quantities of a resource, hence conserving the supply or reserving it for some future date; and (2) by using the resource continually, but with methods permitting its (partial or whole) renewal. Obviously these generalized strategies are geared to particular types of resources. Soil fertility, once lost, cannot be renewed easily or quickly by natural processes, and must be restored by fertilizer which carries a cost in the form of using some exhaustible chemical resource, or in the form of environmental pollution (as in the case of nitrate pollution of water supplies in the Iowa-Illinois corn belt). On the other hand, forage plants are renewable if a suitable grazing regime can be mounted, although even here questions remain as to the long-term effect on soils. Water, of course, can be recycled in various ways and under some circumstances is more easily conserved than soil. While relationships between yield and conservation are very complex the two are interlocked and decision-making should proceed on this assumption - i.e., that yield almost always carries some form of cost in conservation which needs to be considered and managed, i.e., considered as a factor in the making of production decisions.

The model as sketched on Fig. 1 has yield on the vertical axis and conservation on the horizontal. Conservation is measured by the yield one can expect in the future if the strategies followed in the current year are used repeatedly or consistently. The space enclosed by the angle defines a theoretically infinite number of choices between

FIGURE 1

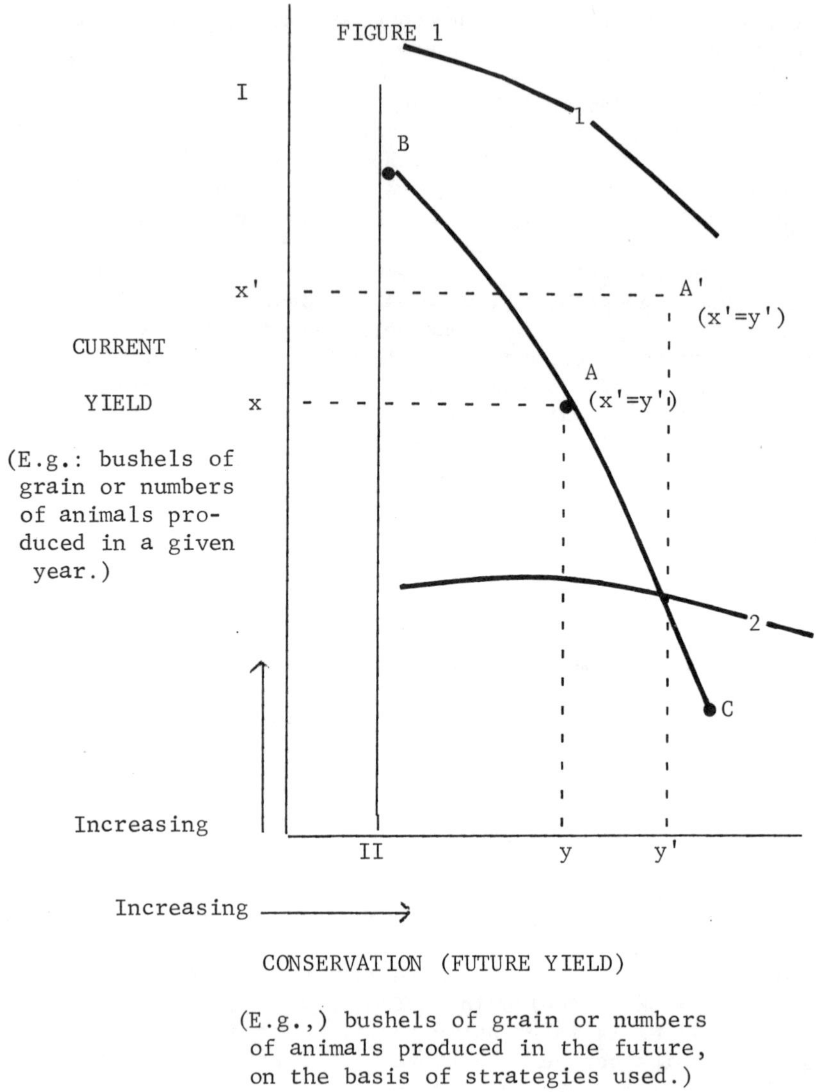

BAC: <u>Optimal</u> Production-possibility curve

A: (x=y): Sustained-yield technique or choice

A': (x'=y'): Theoretical higher-yield sustained-yield position

1 : Resource high-risk abuse curve

2 : Resource preservation curve

strategies which are combinations of current yields and future yields or conservation. This is the basic format of the rational-choice paradigm, and it can be used to show the relationships of any two phenomena - economic technological, attitudinal, behavioral. Theoretically, the choices are infinite, but of course in any real situation the choices, or the technological possibilities involved, will be limited or constrained by a variety of factors.

In some cases, it may be necessary to have the vertical and horizontal axes divided into the same scaled intervals; for others, different scales may be used. However, scales are usually not a problem in the model, and it is possible to represent yield and conservation in different ways. For example, yield can be expressed as pounds, bushels, number of animals, or the financial returns obtained by using particular techniques. Conservation may be defined as pasturage, soil or moisture retention provided by particular yield strategies; or conservation can be defined as the yield that can be sustained over time by such strategies. For example, let us suppose that an operator wishes to obtain the largest possible yield in the shortest possible time; i.e.,point B on the current yield axis of Figure 1. If this strategy is followed repeatedly, it will result in lowered yields, i.e., the low point on the conservation axis which means that (e.g.,) soil fertility has been abused.

On the other hand, if soil abuse is to be avoided, then the operator tries for a lower yield over the short run, knowing that over the long run, the soil will continue to yield at a satisfactory level (point A, or $x = y$). Thus, the conservation scale can measure the amount of yield to be expected, over a period of time, by particular production strategies. This is the measuring system I will refer to below. If the operator follows an equalizing strategy, making current yields signify future yields, he is optimizing his chances by exercising care for his soil. Hence, curves which pass through high points on the current yield axis, signifying short-term, "maximizing" strategies, will curve in toward low points on the conservation axis, meaning that over the long run, yields will decline.[12] If this is to be avoided, the operator would have to make trade-offs between large, quick returns, and the same returns (or something like them) over a longer period of time, or more cogently, a more reliable average yield over this longer period. For example, if the operator were farming on a rental basis, with the conscious intention of making quick returns, he might choose the high yield, low conservation strategy, although soil fertility would be harmed, and the next renter or operator could expect lower yields. If the operator owns the land, and expects to make a living from it for an indefinite period, then the low-initial-yield, long-term conservation and average yield strategy would be better. But if we view the farm as part of a larger resource base which needs to be preserved for the good of a larger population over a longer period, then the conservationist strategy is the best on all counts.

Such choices are fundamental in agricultural strategies the world over, regardless of level of economic and technical development. Strategy decisions of this type are especially important in environments with marginal resources, like arid and semi-arid lands, where crucial problems emerge from the conditions of uncertainty and risk. If marginal environments contain uncertain climates in which adequate moisture supplies are hard to predict, or in which monoculture is subject to price fluctuation out of control of the operator, then the conservationist choices are generally better, since they tend to guarantee a longer average yield with the same resources. That is, they represent management strategies which permit the resources to sustain a yield of product. Consequently agriculturalists who maintain permanent residence in such environments soon develop and maintain conservationist, or "conservative" strategies of management or production, with adjustments for the level of technical sophistication. The curve BAC on Figure 1 represents such a strategic pattern.

So much for fundamentals. Now let us examine the properties of the model in more detail, with the further help of Figure 1. We may begin with BAC, the optimal "production-possibility" curve. As already noted, this curve defines a series of combinations of yield and conservation objectives and techniques with point A representing that combination which will provide a sustained yield indefinitely. We mean "indefinitely" because, as already noted, no agricultural resource is fully renewable, and separate calculations must be made for each real case. Therefore the term, "sustained yield", represents a relative concept, defined by the character of the particular resources, and by the schemes and trade-offs of the producers. However, the generalized meaning of point A is clear enough: it is that point where the techniques in use will, at present levels of information, produce the same amount of yield in the future as today, so long as the yield is measured by units which do not change, e.g., bushels or number of animals. Obviously it is risky to measure the conservation objectives or possibilities in dollars or other transient values, since this will change the meaning of conservation. If prices fall, then point A will, in the future, provide lower returns, even though resources continue to be conserved. Hence there will be temptation to redefine upward the position of a sustained-yield choice point, as indicated by the curve passing through A' (x' = y').

Curve BAC - and all such production-possibility curves - is concave toward the point of origin for a reason: neigher yield nor conservation (which in the present version of the model is also yield) is subject to indefinite increase, as would be the case if the curve were convex. Yields can be increased only to a point: physical constraints begin to operate sooner or later, no matter how sophisticated the techniques. Therefore the amount of yield that can be sustained or averaged by a sustained-yield strategy is likewise constrained. The term, "production possibility curve" thus expresses the amounts of yield which are possible, given the techniques and the resource limitations (e.g.: plants will produce better with more water, but only up to a point, when further applications of water will not produce more yield, or may even damage the plant, and likewise for fertilizers; -in fact such constraints have become critical in the "miracle" strains of rice and other crops developed in recent decades). This concavity of production-possibility curves is extremely important, because often modern agriculture assumes a convexity, at least overtime; i.e., indefinite possibilities of production given changing technology. This view is, of course, influenced by market or profit considerations, or by optimism engendered by scientific advances in agronomy, rather than by the actual state of the resources. However, concavity is also a version of a well-known economics principle: diminishing returns, or the tendency for the marginal or "incremental" gain to fall as successive applications of energy and resources are made. It is interesting that while this principle is embedded in economic analysis, it is often abrogated in agrarian development schemes which assume indefinite increases usually on the assumption that improved technology abrogates the principle.[13]

An example of the operation of production-possibility curves may be drawn from livestock production, where physical constraints can be defined by the concept of "carrying capacity", or the maximum number of animals which can be supported at a given level of nutrition by the forage material produced by a given technique. If the present carrying capacity of a ranch pasture is five animals to the acre for a grazing year (a common figure where native grass pastures are in use in semi-arid environments), and there are 10 acres available, then a yield of 50 animals could be achieved as a sustained-yield condition, a point A. This yield can be obtained on the 10 acres only if the land is grazed in rotation on several adjoining 10-acre plots, so that overuse of any one plot is avoided. Thus, the particular technique (rotational grazing) is a crucial factor. The trade-off between current yield and conservation (future yield) in this example is influenced by the technology, which in turn helps define the position of the production-possibility curve.

It is obvious that resource constraints are quite complex, and that for any production system, several constraints may be at work and interacting differently with the technology. For grain crops, there will be soil fertility, moisture as rainfall, topography, weed cover, and wind speed and direction. For pasture, all these factors may be operative, plus others related to the growth of grasses and the peculiar habits of the animals.

Now, among all possible production-possibility curves, the curve BAC represents the optimal version on Figure 1, since it passes through point A, where future yield (Y) equals, or is defined by, current yield (X). However, we have designated two additional choice and strategy choices: B and C. Point B is a high-yield choice, which minimizes conservation; i.e., seemingly abuses resources so that in the future lower yields can be expected if the choice is repeated often enough. The slippage in the model occurs here: it is possible to "make a killing" once, or a few times, without serious abuse of the soil, and then return to a more modest or sustained-yield strategy. Shrewd farmers and ranchers in arid and semi-arid lands have followed this mixed strategy of operation for a long time, but it contains some behavioral dangers. It tends to establish consumption expectations keyed to the high level of returns, thus making it difficult to revert to conservative strategies.

Point C represents the opposite production choice, an effort to maximize conservation, accepting a small yield. Agricultural producers in market systems would have difficulty understanding this choice, since it seems self-defeating, or at least defines a state of resources. However, this choice may be common as a low-risk strategy to be followed if forecasts suggest drought, heavy winds, or price uncertainties. It may also define special cases of farmers in market systems who are concerned with soil abuse, fertilizer pollution or other problems associated with high-yield, mechanized, chemical-intensive agriculture. In such cases, special ideological incentives must be present in order to induce the operators and their families to accept lower returns, willingly and cheerfully.[14]

It is important to note that BAC is optimal even though it includes choice points related to abuse of resources. It is optimal because by definition, the operator knows where point A is located. Therefore he can, hopefully, return to this point if circumstances have influenced him to make choices like B (which abuse resources), or choices like C, which underutilize resources. Of course, as already suggested, there are some fundamental ambiguities related to the state of knowledge: e.g., point A can be confidently located only relative to existing knowlege. Hence, there is always the possibility that BAC is located too far out, and that point A is really more like point B. Or, that the alternative curve passing through point A' is a delusion, and that A' is not a sustained yield choice. The importance of research in modern high-yield agriculture is thus underlined: we need to know just where our points A are located.

Let us consider some other production-possibility curves. Curve No. 1 describes an arc to the northeast of BAC, on which all production choices exceed the sustained-yield point: i.e., classic high-yield, abusive agriculture which uses all available techniques to influence current yield without regard for future yield, assuming that fertilizer or other technologies will continue to provide suitable yields indefinitely. Curve No. 2 describes an arc south of point A, indicating an extremely conservative strategy which either consciously seeks husbanding or preservation of resources at the expense of yield, or as in some tribal-peasant cases, operates in ignorance of the true potential of the resources. It therefore should be emphasized that the model is not biased toward conservation per se; rather, it allows for choice systems of under use. Thus,

Figure 1 provides three ideal types of production strategy: BAC is the prudent-optimal pattern, allowing for temporarily abusive strategies; No. 1 the imprudent-abusive pattern; No. 2 the traditional-preservationist pattern. Nos. 1 and 2 allow for some movement between yield and conservation, but much less so than the BAC curve.

It should be noted that a farmer, operating on curve No. 1, can assume he is being "conservationist" if he makes a production choice toward the lower or east end of the line. It is true that this point is more conservationist than one higher up, but the importance of the choice is not in the somewhat lower yield sought, but in the basic strategy pattern, which is yield-oriented throughout. Corn-soybean agriculture in the U.S. corn belt provides an example. A farmer may choose a somewhat lower point by growing a little less corn in order to save soil fertility, but this does not materially change the dynamics of the resource-utilization system:-he remains in a yield-intensifying pattern. The reverse is true for No. 2. Here a farmer can try for a higher yield choice, but he remains in a basic hyper-conservationist or underutilization pathway.

Since the choice curves represent a combination of environmental constraints and managerial styles, they can be expected to have a relationship to patterns of selective survival in agrarian systems. An agricultural frontier, like that of the northern Great Plains in the late 19th and early 20th century, will be populated by pioneers who follow many different strategies of production, since knowledge of the carrying capacity of the resources is usually absent. Some pioneers would follow the strategies tested and found adequate in their homelands; others would invent new techniques, and still others would seek to make a quick fortune by doing everything possible without regard for the resources. Some strategies will turn out to be successfully adapted to the resource configuration; others will become casualties of the environmental and economic constraints. In the western Canadian region studies by the writer, over a period of about one generation after settlement, the "get rich quick" type of operator simply vanished, as it became apparent that the resources were too variable and uncertain to support this type of operation on a yearly basis. In effect this meant a replacement of a choice curve of the No. 1 type with one of a BAC type. However, attempts at annual "go for broke" yields appear from time to time with optimism over weather and prices, so no type of managerial strategy is ever ruled out completely. However, these occasional forays are mounted by operators who are fully aware of the risks, and therefore they represent choices of the point B type, rather than choices on a No. 1 type curve.

Another example from the Northern Plains can be drawn from livestock production. Ranchers often display a five- to six-year rhythm in grazing strategies, alternating intensive, "abusive" grazing with periods of underuse, and even herd -reduction, in order to permit the grass to recover. Such ranchers are following a BAC curve strategy, since they are aware of the location of point A, yet if they are studied during the overuse period, they might be erroneously located on a curve like No. 1. Some of the confusion over the concept of "overgrazing" in the agronomy literature probably stems from this temporal alternation of strategies. The opposite would be true for old, first-generation farmers and ranchers whose operations could be defined by the west end of curve No. 2: a naive observer might assume they were simply on the low end of BAC, while in truth they had never utilized their resources optimally.

Turning from semi-arid to tropical resources, in swiddening ("slash and burn") agriculture,[15] the interval between use of a particular field and its recovery and re-use might be as long as ten years, but if a group decided to shorten this interval, and re-use the field before it has had a chance to fully regenerate wild plant cover to replenish soil, the resource system would be running down (i.e., the curve moves from No. 2 to something like No. 1, relatively speaking). The specialist would have to know the cycle precisely in order to determine the pathway of production and conservation strategies.

The relative difference between any and all production choice points can be expressed, of course, with ratios of costs to benefits. However, this is not solely a monetary affair, since the choices themselves involve factors other than economic. Cultural prestige is likely to be involved in choice since styles of production become characteristics of communities. The benefits sought by increases in yield in small-scale entrepreneurial agriculture are related as well to consumption aspirations in the family. Data from the writer's research in western Canada over a decade of observation[16] indicate that 90% of all decisions taken by a sample of about 200 operators of farms and ranches to increase the yield of crops of livestock were taken mainly to satisfy consumption needs of the family members other than sustenance. Over half of the projects undertaken had degradational effects on resources, although in a majority of these cases, the impacts turned out to be reversible; i.e., they were choices which fell along the BAC curve, and by the end of the decade, about one-half of these families had "cut back" (to use the local phrase) in order to permit the resources to recover. Only a minority of about one-sixth of the cases were inveterate "maximizers" who insisted on following something like a No. 1 curve year after year. However, it was also noted that maximizing strategies were common among operators who had recently aquired a farm or ranch, and who felt that it had not been developed to its full potential. In half of the cases of this type observed early in the decade, strategies were already favoring conservationist caution by the end of the period of observation. This led us to characterize the production strategies of this region as, on the whole, conservationist or "conservative", despite the obvious exceptions and alternations.

Such changing patterns of choices suggest that they _are_ governed by rational considerations, however influenced they may be by cultural or personal factors. By "rational" we mean that the operator makes his decisions with a complex texture of precedents and aspirations which include economic factors, but are by no means exclusively devoted to them. Moreover, it was found possible to quantify some of these factors. For example, if an operator decided to forego yield of crop in one field for one year in order to obtain some guarantee of a similar yield a year hence (_e.g._, by saving the field's soil moisture in alternate-year fallowing), he is valuing the moisture and soil capacity in terms of the crop _not_ produced in the present year. This also measures the conservation cost in terms of foregone dollar returns, which, as already mentioned, is a misleading measure because it ignores future gains. To obtain a longer-time measurement, it was found possible to get families to measure the amount of consumption deferred or renounced - not only by dollar costs, but also by cultural benefits, like a decision not to send the son to college, which might be valued in turn by the services the son could provide as a management partner to the father. We found chains of such trade-offs or balancing of costs and benefits to the extent that the whole "cost" of conservationist farming could be expressed by the life style of a family over a generation. Similarly, the reasons for decisions to maximize yield could be determined by examining the specific triggering needs or wants in the family; e.g., the cost of violation of a fallowing routine,[17] and consequent risk of having saved sufficient moisture for a subsequent year, could be a new set of kitchen appliances for the wife. Of course, since this particular population was tied to the market system, there was a strong tendency to measure everything in dollars, but the logic of the decision-making was no different from that in a tribal-peasant case where standards other than money were used.

However, secular changes in the position of the choice curves could be observed taking place over longer periods of time. Regardless of personal or familial grounds for choice, such factors as inflationary trends, gaps between costs and returns, government incentives and disincentives, changing consumption and prestige patterns in the community, changing criteria of performance for agricultural management, and technological changes all had their effects on production strategies.[18] In general, over more than a decade of observation production choices tended to rise, suggesting that conservation was less of a consideration. This was a general response to economic changes and not to improved methods of farming which permitted higher yields at lower conservation costs, although there were some gains in the latter area as well: in particular, better varieties of pasture grasses and improved tillage methods and machinery. However, it is not certain that these long-term changes are permanent. Knowledge of conservation needs and strategies appeared to improve as time passed, regardless of production strategies or yield objectives which seemed to suggest the opposite. It is possible that there are long-term cycles of management as well as the short-term alternations already mentioned. Agriculture in these constrained environments must obey certain rules of production in order to sustain yields over time, hence the possibilities of making fundamental gains (i.e., moving to point A') are, if anything, becoming less certain.

The dynamic of high-yield intensification agrarian systems is to press for more and more production, since the basic governor of the system is to be found in the general culture and its economic and consumption standards. This is as true for the developing nations in arid and semi-arid lands, where difficulties created by undue technological and production pressures on friable resources are already a matter of record. "Desertification" is of course one of the known effects. However, for the developed areas, the yield-increase policy is implemented in the presence of considerable scientific knowledge of possibilities, knowledge which is often ignored or suppressed since it might advise lower performance. This is the situation diagrammed on Figure 1 by the alternative optimal curve passing through point A'. As previously noted, it is important to know whether this curve represents a real gain in sustained yield, or whether it is a product of illusion or deception. The latter is common enough in modern society, with its tendency to make an absolute value out of prosperity and economic gain, and the corresponding tendency to rationalize resource abuse or pollution by some ideology of progress. This is often accomplished by external subsidies to production which artificially reduce the cost of production and likewise, the cost of diminishing resources. This technique has been used in modern agriculture with respect to fossil fuels: when gasoline prices increase for the farmer, tax alleviation or cash subsidies may reduce them, to permit the farmer to continue to produce at an optimal cost without paying a price for the fuel in proportion to the diminishing supply. This also affects soil and water since the farmer will tend to measure the costs of these resources in terms of the subsidized fuel charge. A similar effect can be noted in tribal-peasant production in arid lands where subsidies in the form of food relief and other things may stabilize a population on a resource base which is really not sufficient to continue to produce at the desired level. The strategy seeks to avoid the known difficulty of coming up with ingenious alternatives, which often require complex operations like resettlement, or radical institutional change.

There are some implications here with respect to risk perception and adjustment.[19] In resource conservation, states of risk appear to cut both ways: if farmers fear to risk the loss of income opportunities, they may overutilize resources, or fail to develop them properly. On the other hand, if the fear of resource abuse is strong, then resources may be treated gently or underutilized. It would appear that small economic scale creates greater perception of risk of the first order, while an increase in scale may generate the latter. Studies by the writer of the use of resources by Hutterian Brethren indicate that these responses are cyclical. In the early years of a colony, while it is struggling to obtain sufficient income to support the population and also pay off the debts incurred in establishment, resources will be pressed to the limit and intensive development measures will be delayed in favor of yield maximization. As the colony establishes itself financially, these exploitative practices change toward their opposite: inferior land is taken out of cultivation and built up with recommended techniques; herds may be reduced in size to permit pasture to recover, or newly-sown pastures to mature. Colonies can move from highly abusive regimes to strict conservationist regimes in a matter of two or three years - the same sequence we noted earlier for individual operators, although the scale of the operation will help determine the frequency and magnitude of such changes.

It is obvious that the perception of risk is influenced by many factors other than the state of the business. The need for income in the case of farmers of smaller scale is generated by social pressures from the family, the wider kin and neighborhood group, and many other sources. Frank Cancian (1967) has noted the relationship of risk perception to income levels and socioeconomic stratification, finding that among farmers the "middle class" segment is more conservative, more husbanding of resources than the strata with greater wealth. Thus Hutterites behave like Cancian's "wealthy" farmers (which they are, collectively if not individually) in the early stages of the colony farm, but revert to conservative, resource-husbanding operators in later stages when funds are being saved to permit a colony division and the financing of a daughter group.

These constant movements along production-possibility curves all appear to be characteristic of societies whose production tends to be governed by external forces. As I shall note in more detail later, societies without these inputs tend to develop stable systems, such as isolated tribal or tribal-peasant peoples, an increasingly rare type. The distinctive feature of modern agroecology is that while the operators to an increasing extent know where point A is really located, they are forced, or induced, to repeatedly choose point B as a guide to production objectives and strategies. This is, in general, the basic problem of the human use of resources: <u>why do we choose strategies which exceed prudent limits when we know better</u>?

Another example of an external input influencing production strategies is the export trade in agricultural commodities. Rubber in Southeast Asia, sardines in the Pacific littoral of South America, bananas in Central America, poppies in Asia Minor, and the like have all been analyzed as exploitative devices which the Western or developed world has encouraged to satisfy its own needs, but which have distorted indigenous agricultural regimes, made them vulnerable to monocultural economic fluctuation, or have exhausted resources.[20] Often these crops are produced to supply needed cash to meet rising cultural expectations in the national population (and as often to line the pockets of elites who have no real attachment to their country). These cases simply reinforce the point: as human institutions become intertwined, the growth-oriented forces in human behavior are given added opportunity, and the constraint lines and choice points increasingly tend to feature a withdrawal to the left, or west, on the conservation axis. One of the difficulties with all this is that often the production systems are set in motion before sufficient information on the potential risks can be gathered.

Another factor likely to influence the conception of resource constraint or obedience to it is the particular form of social organization and form of resource tenure. People like the Hutterites of North America, believe that the combination of communal ownership and large scale production facilitates trade-off decisions which optimize both yield and conservation.[21] Collective-use schemes of land and water use in the West also commonly encourage more conservationist practices than do small-scale individual ownership of the same resources. In general, cooperative-collective methods tend to diffuse responsibility over a larger group, and this results in cautious investments. Curiosly, this is viewed as undesirable in the yield- and business-oriented literature or traditional agricultural economics, yet from the standpoint of the conservation goal, it becomes a desirable constraint or control factor. It is also a common pattern in peasant societies.

Agrarian Development

Much of the literature on agrarian development is characterized by assumptions that indigenous, non-powered agricultural technology and methods of cultivation were either abusive of resources or were underutilizing them.[22] Hence development programs involving "rationalization" of procedures and the introduction of high-energy forms of production were justified on the basis of "progress". These symbolic ideas continue to underlie much agrarian development, despite a growing awareness of a more complex situation. First, as already mentioned, some of the traditional methods can turn abusive if some of the factors governing the rate of production or desired yield change. For example, if the interval of fallow in swidden is shortened, caused by increasing population, abridgement of customary range by land appropriation, or rising expectations in the population, a degradational cycle begins and the tropical soils may evolve toward plinthite as plant cover diminishes. Secondly, when traditional methods were adjusted to particular territories and population magnitudes, often they were conservationally optimal, not inefficient or abusive. This case is particularly strong now for arid-land tribal pastoralism, which at lower levels of population was found to be about the most ecologically effective means for utilizing these marginal or friable environments.[23] If a model like the proposed rational choice scheme for analyzing agricultural systems had been applied, it is possible that some of the current unsuccessful attempts to irrigate unsuitable land, sedentarize nomads, or cultivate friable soils might have been avoided.

Thus, the model also helps us to understand the process of achieving relative stability, or slow rates of change, in pre-market or tribal systems of resource use in all habitats, but perhaps especially in those with marginal resources. This problem was a major puzzle for anthropologists, who were at a loss to explain why these small groups of people resisted the pan-human proclivity toward growth and change which has characterized the long historical perspective. However, in recent years a number of explanations have been put forward which shed some light on the question. While these differ on many points, they have a common ground in the willingness to view the resource-utilization system as part of a larger system involving technology, the social system, values, and population. Beginning with pre-industrial-urban low-energy technologies, the group soon standardized resource uses around these methods, which adequately supported a given population at minimal cost to the physical resources. To do this required an interwoven social system which served to control the use of resources by significant subgroups. *E.g.*, in tribal pastoralism, severe sanctions prevented an individual herd-owning clan or family from using pasturage more exploitatively than the others; or among foraging nomads, the fear of social conflict arising from increasing population may have inhibited both population growth and resource and technological practices.[24] However, the social cost of this adaptive control was of course a social environment viewed by contemporary people as stultifying. Hence the more innovative, growth-oriented modern societies, with problems of political equity in the foreground, must evolve formal or legal controls for resource use, and cannot rely on internal systemic controls visible in the low-energy societies.[25]

Obviously some traditional agricultural systems are less vulnerable to culturally-induced urges to increase yields at the expense of conservation than others. Perhaps the least vulnerable is wet-rice cultivation, which is based on a permanent controlled supply of water.[26] Expansion of yield was accomplished in the past by simply increasing the supply and control over the water. The recent introduction of super-high-yield strains of rice has changed the system, since these generally require fertilizer, which introduces a new factor and hence may disturb the balance.[27] This is, moreover, accompanied by rising cultural expectations which reinforce the social demand for higher yields. Hence the traditional system evolves toward a typical contemporary agricultural system in which the resource constraints are changed; or, the <u>illusion</u> is given that they are changed without potential damage to the system. Cycles of degradation and recovery no doubt will follow; whether these can be tolerated remains to be seen.

In desert lands, the climatic factor is much more significant as an input than in regions with more humidity and more equable temperature regimes. In desert conditions, all resources parameters approach marginal tolerances for any sort of yield in agriculture or animal husbandry, although food-gathering groups such as the Shoshone of the Great Basin did very well on seasonal wild products at the cost of very low population magnitudes.[28] Thus even slight changes in population size or migration, as occurred in the Sahel, are capable of contributing to loss of plant cover, soil drifting, and generally what the United States calls "desertification", especially when these are accompanied by lowered resource output levels, such as drought. Any external input such as raised expectations or nationalist sentiments will have accelerated impact, presumably measurable in terms of altered resource procedures.

As the analysis proceeds in this direction, general systems theory begins to enter the field, and the model presented here makes its transition to a more comprehensive study of feedbacks and input-output functions. However, at this point the analyst would hopefully have arrived at fairly precise expressions of the behavioral components involved in the system, and hence the systems model will include the key criteria of human adaptive procedures and processes.

END NOTES

1. An appraisal of rational-choice theory in the social sciences is found in Heath 1976. An appraisal of systems theory in relationship to resource practices and environmental effects is: Man and the Biosphere (MAB) Program 1972. The application of ecosystem theory (as contrasted to general systems approaches) to resource practices is dealt with by Van Dyne 1969. The two systems approaches to the problem are quite different, although there are basic methodological similarities with respect to converting human-factors data into quantitative components.

2. The standard format is the research monograph on a particular tribal society or archeological site. As an example of the former, see Netting 1968. An archeological publication utilizing systems concepts in handling uses of environmental microhabitats is Fitzhugh 1972. In another direction, anthropologists and archeologists have become increasingly concerned with Malthusian questions concerning the relationships among population, subsistence technology, and physical resources in tribal contexts. For a theoretical statement, see Zubrow 1975. Much of the research is inspired by Boserup 1965. An anthropological critique of this book is found in Bronson 1972.

3. E.g.: Bennett 1976a, especially Chapters 3, 4, 8, 9 and Buckley 1967.

4. For an appreciation of the generally neglected factor of time in economic theory and analysis, see Schelling (ed.) 1973, especially the essay by Roberts, p. 646. Studies of the influence of time in human behavior are found in Fraser and Lawrence 1972; 1975. Adaptive behavior in humans, with its articulated sense of purpose and accomplishment, _creates_ time. For an appreciation of this psychological or behavioral basis of time in human behavior, see the papers in Coehlo _et al_. 1976. An appraisal of adaptation as a key concept in anthropology is found in Bennett 1976b.

5. For accounts of the historical change process in early _Homo sapiens_ populations, see the following: Alland 1967; Adams 1966; Flannery 1972a; 1972b.

6. For an introduction to the problems of "desertification", see Berry _et al_. 1975. Experimental studies showing the effects of various forms of cropping and grazing on desert habitats can be found in Novikoff _et al_. 1976.

7. For some data and bibliography on the impact of tribal humans on the physical environment, see Bennett 1976a, pp 78; 80; 151.

8. As Garret Hardin has phrased it, the problem is one of how to achieve control without undue coercion (Hardin 1972). It is possible, however, that this is a typicallyAmerican problem, due to the considerable latitude given the American individual in resource exploitation. The issue is fundamental in the history of the conservation movement, where the standard approach was to blame the "interests" for wasting resources in the face of the "public's" desire to conserve them (e.g., Pinchot 1910, pp.135 ff.); and to advocate conservation as a useful practice or moral position, but to avoid any hint of disciplining the public to reduce living levels--a mechanism visible today as politicians struggle with the necessity to reduce energy consumption. Another way out of the uncomfortable necessity of recommending discipline or sanctions is to theorize about how the market can control exploitative use of resources (e.g., Heal 1975).

9. E.g., MacKrimmon & Toda 1969. For brief explanations of the development of the theory beyond consumption economics and into political science, sociology, and social psychology, see Heath 1976, especially Chapters 1, 2; and notes below:

10. For brief accounts of various facets of this controversy, see Firth (ed.) 1967.

11. Calculations derived from economic models of rational choice are used in the treatment of resource problems by economists, particularly in the field of pricing, public-goods analysis, and others. For examples, see Krutilla & Fisher 1975; and Dorfman (ed.) 1972. The key procedure, as it has developed over the past quarter-century, requires the determination of monetary values for environmental factors, and for any human activity which may be affected by technological change or investment by public or private agencies. Different procedures are required for different kinds of "goods": the valuation of wilderness is a different procedure than that used for assessing the benefits or costs of a new highway. A device used particularly in the determination of costs and benefits associated with natural resources is the "social discount rate"; essentially a valuation of the resource different from the current marginal productivity of capital in order to allow for future resource losses that might be incurred in the development project. One of

the very few detailed analyses of resource conservation (in this case salmon) which handles behavioral incentives systematically from an economic standpoint, is Crutchfield & Pontecorvo 1969. The authors propose that biological conservation methods based on limiting the amount of the Pacific salmon catch will fail, since salmon are an economic good. They propose controlling the numbers of fishermen, and improving the quality of their fishing gear in order to improve efficiency of yield, hence conserving supply of fish. Rational-choice analysis underlies their argument which has certain loopholes, such as the political problem associated with restricting the number of fishing boats and fishermen.

12. It should be emphasized that choice-point "B" presupposes a degree of conservation. Most resource utilization systems recognize the need for some conservation, in order to maintain continuity of operation, although "get rich quick", and "use up and get out" modes of resource-use have always been with us. However, the point is not that the majority of resource-utilization systems are not conservationist, but that they tend to favor yield over conservation, which means that the supply of the resource will be exhausted or depleted sooner than in systems where conservation is given more weight. Discussions of resource conservation like that of Barnett and Morse 1963 seem to use economic standards of resource renewability exclusively, holding that any development of a resource is an economic gain for posterity, and not a loss, even though supply may be reduced. However, this standard ignores technological change, already noted, and also ignores the losses incurred by requirements to reduce the use of a resource, like oil, when the costs rise and the supply has to be rationed. Consumption builds up to a particular level of use, and then if changes have to occur, serious losses and dislocations are suffered, with incalculable political and social consequences. There is, moreover, the problem of resources without clear economic use--a category of increasing importance as ecosystems and genetic material are destroyed by human expansion into habitats, thereby reducing flexibility, alternative use choices, and important genetic variability needed for improving domestic races of plants and animals. See Ehrenfeld 1976 and Felger, this volume.

13. The dynamics of the human system of resource-utilization in agriculture, as contrasted to the style found for plants and animals, is summarized in Odum 1969. Odum's point is that human systems aim toward persistent

imbalance, and production in excess of natural energy supply, whereas in natural communities, the direction is toward homeostatic balance between various organisms so that the avilable energy supply is held relatively stable, only converted and reconverted.

14. An example can be found in the Illinois conservationist or "organic" farmers studies by the Washington University Center for the Biology of Natural Systems. At the time of writing, only one journal article has appeared on this study (Klepper et al., 1977) concerning comparisons on returns and costs for the organic group with farmers who use chemical fertilizers and pesticides. No data on absolute dollar returns and consumption levels--all things which would influence choices of production strategies from the standpoint of the present paper--are given. This research is in its earlier phases and more light can be expected on choices and decisions related to resource conservation.

15. For a brief introduction to the ecological and cultural problems associated with swiddening (also called "shifting cultivation", "slash and burn", "bush-fallow", etc.), see Bennett 1973. This piece also contains bibliography to date of publication. An important issue concerns the extent to which swiddening methods are less abusive of friable tropical resources than intensive sedentary systems. A critique of anthropological assumptions concerning the less destructive impact and swiddening is found in Street 1969. This piece does not settle the issue, however; the impact is a complex affair, involving many variables not yet fully analyzed.

16. In preparation: Bennett, Of Time and Enterprise: Agricultural Management as an Adaptive Process. For earlier versions of some of these data see Bennett 1971 and 1977.

17. The following technique as practiced in North American semi-arid grain agriculture is based on the need to accumulate moisture in the soil to provide sufficient moisture for crops where rainfall is marginal. By not planting grain in a field for one year, and preventing weed growth and maintaining a pulverized surface, moisture from rain and snowfall during the year preceding the next crop season may provide enough to ensure crop growth. Actually in dryer regions of the Great Plains the strategy operates on a three-year cycle, as indicated by the following: the moisture used to produce a grain crop in a given year will be derived from moisture which fell on the

field after harvesting in the fall and winter of, say, 1970; plus the moisture saved during the entire fallow year of 1971, plus the early-spring moisture-fall before planting in the spring of 1972. By resting fields alternately in this manner, soil fertility is also conserved to some extent. With care and proper tillage methods, plus careful use of fertilizer, or even varying crops (e.g., pasture grasses for certain periods, then back to grain), the strategy can come close to a sustained-yield program.

18. Studies of expected crop yield from the standpoint of predicted responses of farmers to incentives and disincentives to modify planting schedules administered by the market or in government programs, have tended to show that behavior can be predicted fairly well if the motivating factors are geared to effective periodicities in the production and marketing systems. An example of research on this problem is found in Sahi & Craddock 1975. This study, done with recursive programming methods, attempted to predict actual acreages planted by farmers for a variety of grain and specialty crops, in response to government incentives, market conditions, precipitation, and the previous year's acreage. "Flexibility indexes" were computed for each crop denoting the extent to which the farmers were willing to alter acreages over the preceding year. Results showed that the predicted acreages were close to the actual acreages for wheat, oats, rye, and summer-fallow, but there was considerable disparity for barley, flax and rape. My data on management behavior among Canadian Plains farmers based on observational and participation research indicate that the three crops which departed from predicted averages are not planted on the basis of single-year rhythms, but are conceived as occasional crops, to be planted at irregular intervals, or at fairly regular off-year intervals, to provide special cash bonuses, or as substitutes for the major cash crops due to needs or incentives derived from local social or economic patterns. Farmers do not view them as subject to management decisions determined by "macro" incentive factors used as variables in the cited paper.

19. For discussions of uncertainty and risk in agricultural production by economists, see Just 1954.

20. For studies of resource extraction in third-world countries, see Tanzer 1969; Seidman (ed.) 1975; Beckford 1972.

21. For a description of Hutterian agricultural and management practices, see Bennett 1967. Conservationist

aspects of Hutterian agricultural are commented on in Bennett 1977.

22. An example, Natural Resources of Developing Countries, Dept. of Economic and Social Affairs, United Nations, N.Y., 1970 (Doc. # E.70.II.B.2). The definition of "natural resource" used in this publication is simply any environmental factor which appears to be "particularly valuable to human welfare, either immediately or in the foreseeable future" (p.5). The report goes on to take the position that most developing countries do not sufficiently exploit their resources to increase their wealth, and that "careless" use of resources by traditional agriculture and mining methods were abusive. However, nowhere in the report is there evidence of awareness that traditional methods of use might be less harmful than the modern methods advocated. Several articles in the present volume demonstrate this point.

23. For discussions of tribal-peasant pastoralism and its ecological impacts, see: Man and the Biosphere (MAB) 1975; Irons & Dyson-Hudson 1972; Leeds & Vayda 1965; Brown 1971.

24. The work has featured foraging (i.e., hunting-gathering-fishing) subsistence adaptations. See Lee 1972; and Woodburn 1972.

25. For brief discussions of how control over resource use emerges as a social problem when the boundaries of local or indigenous social systems are breached by communication and exploitation, see Bennett, 1967A, pp. 52-3; 183; 186-93; 292 ff.

26. For a classic description of the ecology and cultural aspects of wet-rice cultivation, see Geertz 1963.

27. For descriptions of the "green revolution", see Griffin 1975.

28. The classic anthropological analysis of the relationship of subsistence ecology to population among the Amerinds of the Great Basin is Steward 1938.

REFERENCES

Adams, R.C. McC., The Evolution of Urban Society. University of Chicago Press. 1966

Alland, Alex, Evolution and Human Behavior. New York: Garden City Press. 1967

Barnett, H., and Morse, C., Scarcity and Growth. Baltimore: Johns Hopkins Press. 1963

Barth, Frederic, "The Analytical Importance of Transaction", in Models of Social Organization. Occasional Paper No. 23, Royal Anthro. Inst., Glasgow. 1966

Beckford, G., Persistent Poverty: Underdevelopment in the Plantation Economics. New York: Oxford University Press. 1972

Bennett, John W., Hutterian Brethern: The Agricultural Economy of a Communal People. Stanford: Stanford University Press. 1967

__________ Northern Plainsmen, Chicago, Aldine Publ. Co., and AHM Publ. Co. 1971, 1977

__________ "Ecosystemic Effects of Extensive Agriculture", in Annual Review of Anthropology: 2; 36-43 1973

__________ The Ecological Transition. London and New York, Pergamon Publ. Co. 1967A

__________ "Anticipation, Adaptation, and the Concept of Culture in Anthropology". Science: 192 847-53.

__________ Of Time and the Enterprise: Agricultural Management as an Adaptive Process. (in preparation). n.d.

Berry, L. et al., The Sahel: Ecological Approaches to Land Use. Man and the Biosphere (MAB) Program, Technical Notes. 1975

Blair, Peter M., Exchange and Power in Social Life. New York: John Wiley. 1964

Boserup, Ester, The Conditions of Agricultural Growth. Chicago: Aldine Publ. Co. 1965

Bronson, Bennet, "Farm Labor and the Evolution of Food Production", Spooner, 1972

Brown, L.H., "The Biology of Pastoral Man as a Factor in Conservation". Biological Conservation: 3; 93-100. 1972

Buckley, Walter, Sociology and Modern Systems Theory. Englewood Cliffs: Prentice Hall. 1967

Cancian, Frank, "Stratification and Risk Taking". American Sociological Review: 32; 912-27. 1967

Coehlo, George V. et al., Coping and Adapting. New York Basic Books. 1976

Crutchfield, J.A., and Pontecorvo, G., The Pacific Salmon Fisheries: A Study of Irrational Conservation. Baltimore: Johns Hopkins Press. 1969

Dorfman, R. (ed.), Economics of the Environment. New York: Norton. 1972

Dorner, Peter (ed.), Cooperative and Commune: Group Farming in the Economic Development of Agriculture. Madison: University of Wisconsin Press. 1977

Downs, A., An Economic Theory of Democracy. New York: Harper & Row. 1957

Ehrenfeld, D.W., "The Conservation of Non-Resources", American Scientist:. 64; 648-56 1976

Firey, Walter, Men, Mind and Land, Glencoe: The Free Press. 1959

Firth, Raymond (ed.), Themes in Economic Anthropology. London; Tavistock. 1967

Fitzhugh, William W., Environmental Archeology and Cultural Systems in Hamilton Inlet, Labrador. Smithsonian Contribution to Anthropology, No. 16 Washington, D.C.

Flannery, Kent V., "The Origins of the Village as a Settlement Type", in P.J. Ucko & Others, 1972a

______________ "Evolutionary Trends in Social Exchange and Interaction", in E. Wilmsen, 1972b

Fraser, J.T., and Lawrence, N., The Study of Time: Proceedings of the International Society for the Study of Time. New York: Spring-Verlag. Vol. I, 1972: Vol. II, 1975. 1972, 1975

Geertz, Clifford, Agricultural Involution. Berkeley: University of California Press. 1963

Griffin, K., The Political Economy of Agrarian Change: An Essay on the Green Revolution. Cambridge: Harvard University Press. 1974

Hardin, Garrett, "Tragedy of the Commons". Science: 162 1243-48. 1968

______________ Exploring New Ethics for Survival, New York: Viking Press. 1972

Heady, Earl O. et al., Economic Instability and Choices Involving Income and Risk in Primary or Crop Production. Ames, Iowa Agricultural Experiment Station Research Bulletin 404. 1954

Heal, G., "Economic Aspects of Natural Resource Depletion", in Pearce & Rose, 1975. 1975

Heath, Anthony, Rational Choice and Social Exchange. Cambridge University Press. 1976

Homans, George C., Social Behavior: Its Elementary Forms. New York: Harcourt Brace. 1961

Irons, W. and Dyson-Hudson, W., Perspectives on Nomadism. Leiden, 1972

Just, R.E., "Risk Response Models and Their Use in Agricultural Policy Evaluation". American Journal of Agricultural Economics: 57; 836-43. 1975

Kahn, M., The Economy of the Green Revolution in Pakistan. New York: Praeger. 1975

Klepper, Robert, et al., "Economic Performance and Energy Intensiveness on Organic and Conventional Farms in the Corn Belt: A Preliminary Analysis". American Journal of Agricultural Economics: 59; 1-12. 1977

Krutilla, J. and Fisher, A., The Economics of Natural Environments. Baltimore: Johns Hopkins Press. 1975

Lee, Richard, "Work Effort,Group Structure, and Land Use in Contemporary Hunter-Gatherers", in Ucko, 1972. 1972

Leeds, A. and Vayda, A.P., Man, Culture and Animals. Washington: American Association for the Advancement of Science. 1965

MAB, Man and the Biosphere (MAB) Program, Report, Series No. 2, Expert Panel on the Role of Systems Analysis and Modelling Approaches. Paris, UNESCO. 1972

____ Man and the Biosphere (MAB) Program,Report Series No. 25: International Working Group on Project 3: Impact of Human Activities on Grazing Lands. Paris, UNESCO. 1975

MacKimmon, K.R. and Roda, M., "The Experimental Determination of Indifference Curves". Review of Economic Studies: 36; 433-52 1969

Mauss, Marcel, The Gift. (trans. by I. Gunnison). London: Cohen & West. (original publication in French, 1925). 1925

Netting, Robert M., Hill Farmers of Nigeria. University of Washington Press. 1968

Novikoff, G. et al., Desert Biome: Joint MAB and U.S. IBP: Tunisian Presaharan Project: Progress Reports for the Years 1971-76. Logan, Utah State University, 1976. 1976

Odum, Eugene P. "The Strategy of Ecosystem Development", Science: 164: 262-70. 1969

Olson, Maneur, The Logic of Collective Action. Cambridge: Harvard University Press. 1965

Pearce, D.W. and Rose, J., The Economics of Natural Resource Depletion. London: Macmillan Press. 1975

Pinchot, G., The Fight for Conservation. New York: Doubleday Page. 1910

Roberts, Marc J., "On Time": essay in Schelling, 1973.

Sahi, R.K. and Craddock, W.J., "Estimating Crop Acreages in Prairie Provinces", Canadian Journal of Agricultural Economics: 23; 1-15. 1975

Schelling, R.C. (ed.), "Symposium on Time on Economic Life", Quarterly Journal of Economics: 57; 627-668. 1973

Seidman, A. (ed.), Natural Resources and National Welfare: The Case of Copper. New York: Praeger. 1975

Spooner, Brian (ed.), Population Growth. Cambridge: MIT Press. 1972

Steward, Julian, "The Economic and Social Basis of Primitive Bands", in Essays in Honor of A.L. Kroeber, Berkeley: University of California Press. 1936

Street, J.M., "An Evaluation of the Concept of Carrying Capacity". Professional Geographer: 21; 104-7 1969

Tanzer, M., The Political Economy of Oil and the Underdeveloped Countries. Boston: Beacon Press. 1969

Thibaut, J.W. and Kelley, H.H., The Social Psychology of Groups. New York: John Wiley. 1959

Ucko, P.J. et al., Man, Settlement and Urbanism. London: Duckworth, and Cambridge: Schenkman Publ. Co. 1972

United Nations, Natural Resources of Developing Countries, Dept. of Economic and Social Affairs, United Nations, N.Y. 1970

Van Dyne, George, The Ecosystem Concept in Natural Resource Management. New York: Academic Press. 1969

Wilmsen, E. (ed.), Social Exchange and Interaction, Anthropological Papers, Museum of Anthropology, University of Michigan: Ann Arbor. 1972.

Woodburn, James, "Ecology, Nomadic Movement, and the Composition of the Local Group among Hunters and Gatherers", in Ucko, 1972. 1972

Zubrow, Ezra, B.W., Prehistoric Carrying Capacity. Menlo Park: Cummings Publ. Co. 1975

Nicholson and Reining were asked to provide oral discussions of the papers at the symposium in Denver. As so often happens, they did not receive copies of all papers in time to prepare their remarks. Nicholson's contribution derived from customary procedure in some other disciplines - she wrote a completely independent short statement, to which she appended some reactions as she listened to the presentations. Since her field is climatology, her views are an especially welcome addition here, for they provide a cautioning note to those anthropologists who may still not fully appreciate the importance climatic factors may have had in influencing the course of history. Her long-range perspective reinforces that of many of the earlier papers, but especially that of Moseley and Bedoian, and urges us not to jump to conclusions on the basis of immediate observations.

7

Drought versus Desertification: The Case of the Sahel

Sharon E. Nicholson

In considering the recent dry episodes which have plagued numerous and widespread regions of the world, a distinction between drought and desertification is too seldom made. The former, which can be defined in many ways, is essentially a transitory period in which a deficiency of rainfall produces hardships on a population. "Desertification" suggests a trend toward a drier set of mean conditions, including a change toward vegetation in equilibrium with the new situation. Sometimes the term refers to a trend partially caused by human factors or to one caused by natural factors, but reversible only by conscious human efforts.

It is difficult to assess on climatic evidence alone whether a process of desertification is taking place in a particular region, since we are limited in establishing mathematically defined trends to the short period of time for which records have been kept, the past 100 years or less. A trend over a fifty-year period commencing in an anomalous wet episode will differ considerably from the one determined when twenty very dry years are added to the beginning of the series. Furthermore, the inherent "noise" in the series, the irregular year-to-year changes of weather, is likely to be of greater magnitude than the trend, thereby obscuring actual trends or creating apparent ones.

The contrast between the dry conditions in the Sahel in the late 1960's and 1970's and the so-called normal rainfall of the early and mid-twentieth century is often interpreted as a trend toward increasing aridity, i.e., desertification. However, the whole concept of "normal" rainfall in such semi-arid regions is a statistical fiction (Katz and Glantz, 1977) because natural variability there is very high. Annual rainfall totals most often fall below the normal, or mean, value, which is inflated by a few years with rainfall far in excess of normal. In regions with very low rainfall, a significant change in the annual total can result from only a slight change in the climatic or weather regime - perhaps one or two additional falls of rain. Thus, from a climatic point of view the recent rainfall anomaly in the Sahel appears to have been simply a drought - a passing dry episode which has already given way to wetter, more "normal" conditions. However, a look at longer-term fluctuations of the region's climate, using the methodology of historical climatology, helps us further assess whether we are really witnessing drought or desertification.

From the mid-sixteenth to some time in the eighteenth century, the entire semi-arid region south of the Sahara experienced higher precipitation than during most of the present century (Nicholson, 1976). There is evidence that these conditions covered the entire east-west extent of the region. Adanson's eighteenth century description of the area along the Senegal River (bordering Senegal and Mauritania) contrasts sharply with the present landscape near and north of the river, an almost vegetationless expanse of sand. Although he visited the area during a period of severe drought about 1750, he nevertheless mentions a thick wood bordering the island of Sor near the river's mouth at St. Louis and a forest consisting of tamarisks, redgum trees, and several sorts of thorny acacia. The French planted several species there which were said to have thrived "amazingly" and included oranges, lemons, limes, figs, pomegrantes, guavas, pineapples, papaws, sour-sops, and cashews. Potatoes were said to be abundant in the wet, marshy lands near Podor. Maps from the early eighteenth century show lakes in Senegal and Mauritania which no longer exist today, as well as forests in southern Mauritania, where now are found only dunes. Considering only the contrast between this period and the twentieth century, a long-term trend of desertification might be suggested.

From about 1738-56 a drought as severe as the one in 1968-73 ravaged the entire Sahel, killing half the population of Tombouctoo (estimated to have been then about 100,000) and other nearby areas. Observers at that time would probably have said, as we do today, that the whole region was drying up. Not unlike today, some blamed the situation on human factors. Moroccan invaders, who conquered the Songhai state of the Niger Bend region in 1591, were held responsible for a general degradation of conditions, including the 1738-56 drought which, however, was undoubtedly climatically determined (Nicholson, 1976).

Did this drought of the mid-eighteenth century start an irreversible trend toward desertification which the Sahel is still experiencing? The answer appears to be, "No". There are several later examples of wetter, more "prosperous" periods, as well as recurrences of severe drought. One wetter episode occurred as recently as the late 1800's, following a severe drought which lasted during most of the first half of that century. Again, these conditions were evidenced across the entire east-west extent of the Sahel and also along the temperate margin of the Sahara. The lowland area of the Niger Bend, about 80 km west of Tombouctoo, which had been devastated by drought in the mid-eighteenth century and ravaged again within the last decade, became in the 1880's and 1890's the most important wheat-producing area of West Africa (Plote, 1974).

These conditions ended abruptly after about 1895, when decreased rainfall and poor floods of the Niger ended all agricultural activity in the region within a few years (Plote, 1974). Increasing aridity throughout the entire Sahel culminated in a drought between 1910 and 1920 as severe as that of 1968-73. After the 1920's the good rains returned to the Sahelian region, though never in as great abundance as in the late 1800's nor in parts of the sixteenth through eighteenth centuries (Nicholson, 1976).

What can these long-term changes tell us about drought and desertification in the Sahel and, by analogy, other arid and semi-arid regions? First of all, over the past centuries there have been frequent and rapid fluctuations in the Sahelian climate, rather than a long-term trend toward increased aridity. Drought is an ever-present climatic element in the Sahel and other regions, but what is relevant is the changing frequency of its occurrence. Such a change in the reliability, i.e., variability, of the rains can be culturally more important than a change in the mean amount of rainfall. There is some evidence (Talbot and Williams, 1976) that such a change in rainfall variability has occurred recently in the Sahel.

The inherent irregular variability of climate and weather also underlines the need for reserved interpretation of climatic trends and leads us to question the usefulness of prognostic trends. Landsberg (cited in Katz and Glantz, 1977) goes so far as to call these "statistical fictions". Climatic trends represent only a small portion of an infinite series and they also, especially in semi-arid regions such as the Sahel, can rapidly reverse at any time. This is apparent not only in the long-term series, but also in the short-term record: the period of well-above-average rainfall in the Sahel in the 1950's and 1960's (Nicholson, 1976) gave way to severe drought in the late 1960's and early 1970's.

Thus, there is no climatic evidence that the recent Sahel drought is part of a long-term process of desertification. On the other hand, we cannot afford to neglect the possibility that people, particularly in their response to drought, may do irreversible damage that may in fact lead to desertification in such delicately balanced ecosystems as the Sahel. We know that it is possible to alter climate, at least on a small scale - urban climates are a good example. There is also increasing evidence that we may be altering climate on a macro- or even global scale (Bryson, 1973, SMIC, 1971, Charney, 1974). Thus, we must recognize drought as an integral part of climate while leaving open the possibility of significant and perhaps overwhelming changes due to human intervention.

Moseley's paper serves to remind us that the environment is not static and that it is important for those studying past, as well as present cultures, to realize that the environmental stress to which peoples are subjected varies through time. A culture must be able to adjust to a fluctuating, rather than a mean set, of environmental conditions, and agricultural practices must allow for such adaptation. Moseley's paper illustrates what may happen when they do not. However, we should not forget that the irrigation works of the Moche Valley may have been severely influenced for better and for worse by the climate. Several of the periods of expanded cultivation occurred during quasi-global climatic episodes in which conditions differed significantly from those of today. In particular, two of these periods were characterized by increased precipitation in tropical Africa and similar fluctuations might be expected in tropical South America, as the regions are influenced by analogous general atmospheric circulation features. The period 500 B.C. to the first few centuries A.D., when the Moche culture was expanding, was wetter than at present in Mauritania and Senegal (western Sahel and southwestern Sahara), and the Nile floods were considerably higher than today. Increasing aridity set in during the first few centuries A.D., when sand deposition again became active in Middle Egypt and the Libyan desert. Perhaps the advance of the sand dunes in the Moche Valley, which resulted in agrarian collapse, had a similar climatic origin.

Another "humid" period from the ninth through thirteenth centuries in Senegal and Mauritania eastward to Niger allowed for the expansion of African "empires" into lands which are presently desert: this same period saw the expansion of the Chimu empire in Ecuador and Peru.

One might also wonder if the widespread climatic change

which occurred around the fifteenth or sixteenth century (often referred as the "Little Ice Age") had an effect on the third instance of agrarian collapse in the Moche Valley. Such a major change would likely have had repercussions in South America; however, in the African tropics, the change was toward increased precipitation.

The points made by Felger concerning (1) greater diversity in subsistence species and (2) farming of crops better adapted for a particular environment are particularly relevant in light of the great natural variability of climate in arid and semi-arid regions. Periods in which rainfall differs from the "normal" are the rule rather than the exception; thus, crops that can adapt to a wide range of conditions should be favored, as opposed to, for example, high-yield crops which produce an abundance under unreliable "normal" conditions, but fail badly with a slight departure from normal rainfall. Similarly, changes in rainfall regimes frequently involve anomalous seasonal distribution of precipitation rather than anomalous annual totals. Diversification of species used for food will provide one way of coping with this and for avoiding those conditions which may lead to irreversible damage in areas under stress.

The various case studies discussed by Moseley, Felger, and others at the symposium illustrate well the interrelationship between people, agriculture, and climate. They point out that the confining and uncontrollable boundary conditions set by nature become more or less restrictive depending on our response to them. Needed are more case studies, such as those presented here, which can establish what response to a particular set of less than optimal natural conditions - climate or land - will create an optimal living environment for those coping with these conditions.

References

Bryson, R. A.
1973, Climatic Modification by Air Pollution, II: *The Sahelian Effect*. Institute for Environmental Studies, University of Wisconsin, Madison, Rep. 9.

Charney, J. G.
1975, Dynamics of Deserts and Drought in the Sahel. *Quar. J. Roy. Met. Soc.*, 101 (428): 193-202.

Katz, R. W. and M. H. Glantz.
1977, Rainfall Statistics, Drought, and Desertification in the Sahel. *In* Desertification: Environmental Degradation

in and Around Arid Lands. M. H. Glantz, ed., Boulder, Colorado, Westview Press.

Nicholson, S. E.
1976, A Climatic Chronology for Africa: Synthesis of Geological, Historical, and Meterological Information and Data. Ph.D. Thesis, Department of Meterology, University of Wisconsin, Madison.

Plote, H.
1974, L' Afrique Sahélienne se Desseche-t-elle? Bur. Rech. Géol. et Minières, Service Géologique National, Orléans, France.

SMIC.
1971, Inadvertent Climate Modification (Report of the Study of Man's Impact on Climate - SMIC). Cambridge, MIT Press.

Talbot, M. R. and M. A. J. Williams.
1976, Dune Erosion and Alluvial Fan Formation at Janjari, Central Niger. unpublished manuscript.

The final set of comments is by Reining. During the Denver session she attempted to summarize, criticize, and extend some of the ideas presented, many of which she only encountered as they were delivered. Her special insights derive not only from a background of ethnographic work in the Sahel, but from recent active participation on the Committee on Arid Lands of the AAAS. As such, she has kept abreast of the work being done not only by anthropologists but by those other scientists who find the sociocultural perspective a valuable complement to their own observations and interpretations.

8

Some Observations on Adaptation to Semi-arid Environments

Priscilla Copeland Reining

The issues discussed in these papers center on the relationship of human activity to the process of desertification. The nature of dry lands management is clearly central to an understanding of these issues and we are indebted to the authors for providing these case studies. Before considering the impressive research results presented here, I would like to make several observations which will put my discussion into context.

The first is that in contemporary populations, those populations following hunting and food gathering as a way of life in arid or semi-arid environments are a tiny percentage of the world's total population. This has not always been the case; in hominid and human evolutionary history, adaptation to semi-arid environments was a very important stage in the development of the human species itself. Studies of hominid fossils discovered in intertropical African sites suggest that development over a three million year period probably took place in open, dry, warm savanna. Under these conditions naturally occurring watering places were vital. Although it is understood that climatic variations occurred in Africa during this time period, the contrast between eastern African plateau and high or low altitude rain forest is important for understanding the contrast between hominid development and non-hominid primate development. Hominid biological and sociocultural adaptation to semi-arid grasslands produced several of the characteristics of our species, including bipedal locomotion, upright posture, and enlarged brain case, improved vision and various behavioral and social accoutrements such as tool making and, we infer, band life. In short, the early human condition has been glossed as an adaptation to semi-arid environments, with later secondary radiation into other environments. The now published review of developments in southern Africa (Science 8 July 1977:

115-126) reiterates the dominant characteristics of that environment and of that development.

This is not the time or place to recount the complex history of secondary adaptation and settlement in many other habitats or environments, an accomplishment which now dominates contemporary human habitation. All of the studies in this volume are concerned with human activity in comparatively recent or contemporary semi-arid and arid environments. However much is new in these adaptations, it is worth recalling that they are built on a base of some of the older strategies for coping with semi-arid environments which were once crucial to the development of the human species.

Secondly, it is useful to note the current estimate of the dry-land population of the world, taken from one of four reviews prepared for the U.N. Conference on Desertification, entitled Population, Society and Desertification (Kates, Johnson, and Haring: 1976). More than six hundred million or fourteen percent of the world's population live in semi-arid lands. The amount of land categorized as semi-arid or arid constitutes one-third of the total land surface in the world.

Thus, dry-lands management involves an attempt to cope with a major portion of the earth's surface and its resources by a significant portion of the world's population. Their successes and problems can be instructive. Considered here are only a few of the many socio-cultural adaptations recent human groups have made to semi-arid environments. For example, the southern Sahel, the border between the true Saharan Desert and the wetter lands to the south in sub-Saharan Africa, is home to some three hundred different groups of people in the 5,000-mile strip between the Atlantic Ocean and the Red Sea/Indian Ocean. There are many different micro-habitats, and there are also different micro-climatic areas. One could take the research results presented in these papers and use them as a basis for further comparative studies in a region such as the Sahel.

Especially appropriate for its methodology in Bedoian's paper on the sedentary Arabs in the northern Sahel--the PreSaharan ecosystem. Even though the subsistence crop is wheat, not the millets and sorghums of the southern Sahel, and the concentration of both farms and animals in single families contrasts with the farming/herding split to the south, his study of crops, animals, marketing, cash labor, and purchase of mechanical services in relation to variation in rainfall and crop yield, is a model which could well be followed with profit in the southern Saharan fringe where marketing and seasonal migration for cash income can also be found. The nature of the adaptation during low, average and double average rainfall is also especially instructive, since extreme variability is a commonly noted feature of low rainfall regimes.

There is a strong tendency among the experts preparing for the U. N. Conference to consider human impact on the environment as the primary force leading to land degradation. To clarify this point, I would like to insert the definition of desertification which is now incorporated into the United Nations Plan of Action, the major document prepared for consideration at the forthcoming conference:

> "Desertification is an aspect of the widespread deterioration of ecosystems under the combined pressures of adverse and fluctuating climate and excessive or ill-advised exploitation. Such exploitation has diminished or destroyed the productive potential of food and fibre-producing ecosystems at a time when increased productivity is needed to support growing populations in quest of development. Important factors in modern society--the struggle for development, population growth and change and the effort to increase food production--interlock in a network of cause and effect. Progress in development, planned population growth, and improvements in food production must therefore be dealt with together. The deterioration of productive ecosystems is an obvious and serious threat to human progress. In general, the quest for ever greater productivity has intensified exploitation and has carried disturbance by man into less productive and more fragile lands. Over-exploitation gives rise to degradation of vegetation, soil and water, the three elements which serve as the foundation for development in many nations. Desertification is a self-accelerating process, feeding on itself, and as it advances, rehabilitation costs rise exponentially. In exceptionally fragile ecosystems, such as those on the desert margins, the loss of productivity through the degradation of soil can easily become irreversible, permanently reducing their capacity to support human life. Action to combat desertification is required urgently before the cost factor rises beyond practical possibility or before the opportunity to act is lost forever." (U.N.: 1977)

If we are to devise adequate anti-desertification measures, we may profit from the research results already available for a variety of different semi-arid environments which enable us to examine the indigenous modes of adaptation for their controls, that is, for the extent to which they conserve the resource base. It is assumed that conservation of the resource base over a defined period (e.g. several centuries) involves "anti-desertification," conscious and/or implicit means of countering tendencies toward land degradation. In addition we also need to consider the difference between indigenous forms of management in dry lands and what is termed in the development community, "good land management," as this is informed by western concepts of range management and rain-fed agriculture.

In the articles in this volume the focus of research and discussion is the nature and functioning of different types of social organizations situated in dry lands. There is considerable difference in their levels of complexity. Each system, however, in its time and place, displays a selection of coping mechanisms, whether involving collecting, herding or irrigation cultivation. These coping mechanisms generally have not led, so far as we can tell, to land degradation.

The first lesson to be learned from these articles is that people can devise means, over time, of utilizing scarce, variable or unpredictable resources in a wide variety of ways under "normal" semi-arid or desert conditions. The second major lesson to be learned is the significance of phenomena--natural, political, and, more recently, technical--that may overwhelm the "normal" adaptive system.

The first of such overwhelming phenomena are natural hazards of rare occurrence--the torrential rains that occur only once in four centuries, the truly prolonged drought that lasts thirty years rather than three, or the dust storm of uncommon magnitude. Some of these natural hazards more or less permanently alter the landscape, such as tectonic movement leading to the epiphenomena of sand dune formation described by Moseley. The means of measuring and interpreting such natural hazards have improved to such an extent that we can also extrapolate to measure earlier events, as Moseley demonstrates.

The second type of phenomenon--political hazards--arises neither from problems posed by the environment nor from the lack of technical solutions. Viewed from the vantage point of these papers, the political systems may either permit or prevent achieved technical solutions from continuing in use. Vidal's entire paper is a testament to this proposition, as is much of that of Lees. Moseley's discussion also shows that Inca policy was at least as much of an obstacle to intensive use of the Moche Valley as the natural hazards he also describes.

Finally, as shown by Lees and Bedonian, technical hazards arise in modernization processes under conditions where neither political nor natural hazards are causative factors but where procedures of technology transfer interfere with adaptation to contemporary conditions.

The lessons to be learned from these papers are potentially useful both to researchers and administrators, who should now distinguish between major and minor fluctuations or hazards, and consider the significance of technical and political factors in addition to the natural ones upon which they have more often concentrated.

References

Kates, Robert; Johnson, Douglas and Haring, Kirsten
1976: Population, Society and Desertification, Clark University, Worcester, Massachusetts.

Klein, Richard
1977: The Ecology of Early Man in Southern Africa. Science 197: 116-126

Secretariat of the United Nations Conference on Desertification
1977: Plan of Action to Combat Desertification. Second preliminary draft, Nairobi.